COMPUTERS IN
CRITICAL CARE AND
PULMONARY MEDICINE
VOLUME 2

A Continuation Order Plan is available for this series. A continuation order will bring delivery of each new volume immediately upon publication. Volumes are billed only upon actual shipment. For further information please contact the publisher.

COMPUTERS IN CRITICAL CARE AND PULMONARY MEDICINE
VOLUME 2

Edited by
O. Prakash
Erasmus University
Rotterdam, The Netherlands

Associate editors:

A. A. Spence
University of Glasgow
Glasgow, United Kingdom

J. P. Payne
Royal College of Surgeons
London, United Kingdom

B. Jonson
University of Lund
Lund, Sweden

S. Nair
Yale University School of Medicine
Connecticut, U.S.A.

PLENUM PRESS • NEW YORK AND LONDON

Proceedings of the Second International Symposium on
Computers in Critical Care and Pulmonary Medicine,
held June 3–6, 1980, in Lund, Sweden

Secretary-general and chairman program committee
B. Jonson, M.D., *Lund, Sweden*

Chairman
O. Prakash, M.D., *Rotterdam, The Netherlands*

Co-chairman
S. Nair, M.D., F.A.C.P., *Norwalk, Connecticut, U.S.A.*

STEERING COMMITTEE
Dr. R. Imbruce, *Bridgeport, U.S.A.*
Dr. J. Osborn, *San Francisco, U.S.A.*
Dr. R. M. Peters, *San Diego, U.S.A.*
Dr. R. M. Gardner, *Utah, U.S.A.*
Dr. B. D. McLees, *Bethesda, U.S.A.*
Dr. I. Staw, *Bridgeport, U.S.A.*

Coordinator
Miss B. Richardson, *Lund, Sweden*

©1982 Plenum Press, New York
Softcover reprint of the hardcover 1st edition 1982

A Division of Plenum Publishing Corporation
233 Spring Street, New York, N.Y. 10013

ISBN 978-1-4615-6706-6   ISBN 978-1-4615-6704-2 (eBook)
DOI 10.1007/978-1-4615-6704-2

PREFACE

This volume, the second in a series on topics in microcomputers in critical care and pulmonary physiology, contains the proceedings of the Second International Symposium on Computers in Critical Care and Pulmonary Medicine, held at the University of Lund in 1980 under the chairmanship of Prof. B. Jonson, M.D., Department of Clinical Physiology, University of Lund, Sweden.

Clinicians and biomedical engineers from many countries participated in a three day deliberation. Of special interest was the introduction of nuclear techniques in pulmonary medicine for the first time in this symposium series.

It is the intention of the steering committee that such meetings should take place on an annual basis in the rapidly changing world of the science and technology of computing in clinical care, in practice and in pulmonary medicine.

Editorial modification of the papers in this volume has been kept to a minimum. Changes have been made to ensure some uniformity in presentation and there has been some alteration of the English to avoid ambiguity, but our intervention has gone no further than that.

It is hoped that the contents of this volume will enable those who are interested in the subject matter to be more aware of research developments occurring in so many different disciplines and so many different centres in America and Europe.

Finally, I would like to thank Miss Bodil Richardson for her organisational and secretarial help. Thanks are also due to Prof. J.P. Payne, Prof. A.A. Spence, Prof. B. Jonson and Prof. S. Nair for their helpful suggestions and assistance in editing this volume.

O. Prakash

CONTENTS

LUNG MODELS

REGIONAL LUNG FUNCTION AND IMAGING

CONTENTS

COMPUTER SYSTEMS IN THE INTENSIVE CARE UNIT

CLINICAL APPLICATIONS OF COMPUTERS IN THE
INTENSIVE CARE UNIT

CHAPTER 1

LUNG MODELS

THE PATHOPHYSIOLOGY OF AIRWAYS DISEASE

Gordon Cumming

Midhurst Medical Research Institute
Midhurst
West Sussex

There is a group of chest diseases in which dyspnoea is associated with limitation of expiratory airflow and it is to the treatment of this impaired airflow that therapy is mainly directed and to which the drug industry devotes much attention.

Implicit in this attitude is the assumption that the characterising feature of the group of diseases is similar and that it is only limitation of expiratory airflow. In addition there is a defect of gas mixing, and the conventional explanation is that it results from a failure of gas distribution by the airways. There is also a further assumption, which is that the process of gas mixing by diffusion within the terminal ventilatory unit is complete within the time of a respiratory cycle.

So we have a unitary hypothesis which explains both airflow limitation and gas mixing by the same mechanism and this represents the conventional wisdom.

It follows from this hypothesis that the more severe the airflow limitation, then the greater must be the defect in gas mixing so that measures of both in a series of patients would show a good correlation. Such a study carried out in my laboratory is shown in Fig. 1. where $FEV_{1.0}$ is used as a measure of airflow limitation and alveolar mixing efficiency for nitrogen is also plotted. As you see the correlation is poor, and the correlation coefficient calculated was 0.06.

This observation denies the conventional unitary hypothesis and suggests that airflow limitation and gas mixing are independent variables. Since this is so a new hypothesis for the genesis of impaired

3

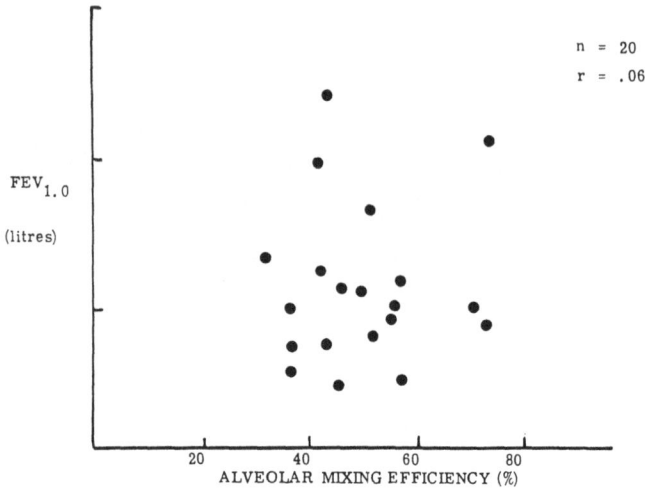

Fig. 1. Correlation between airflow limitation and gas mixing
        efficiency in asthmatic subjects.

mixing is called for, and this hypothesis must not involve the function of large airways.

    It is possible therefore:-

    (a) that it is concerned with the small airways and
    (b) that it is concerned with the assumption made in the uni-
         tary hypothesis that gas mixing by diffusion is complete
         within one respiratory cycle.

    Understanding the process of gas mixing is made easier by con-
sidering the situation which obtains when a subject inspires pure
oxygen and the nitrogen concentration in the expirate is measured,
giving a trace such as is shown in Fig. 2(a) in which expired nitrogen concentration is plotted against expired volume. From such a
trace the slope of the plateau can be measured and an increase in
this slope has been used for clinical diagnostic purposes and the
explanation for its origin has used regional ventilation as its
basis.

    More information can be obtained from the same data if the quan-
tity of nitrogen expired is calculated. Figure 2(b) indicates that
expired flow is measured at the same time as expired concentration.
The two curves are sampled by a computer as indicated in Figure 2(a)
and (b), and each value of concentration is multiplied by each value
of flow, the product indicating the quantity of nitrogen evolved
during the short sampling period.

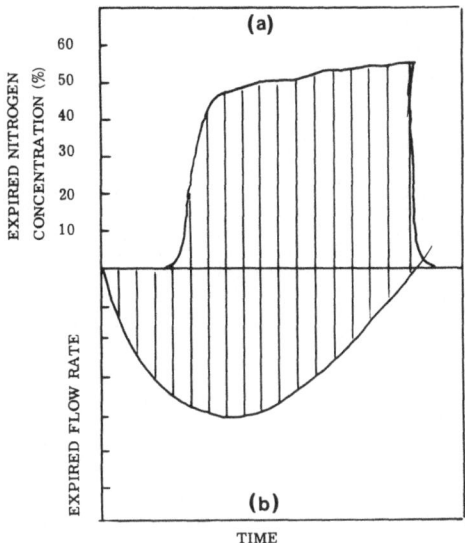

Figs. 2(a) and 2(b)

When all the small quantities of nitrogen are added, the pattern of nitrogen evolution with respect to expired volume may be seen in Figure 2(c). Figure 3 shows the individual data points and Figure 4 the graphic representation, and from this we can draw several conclusions - firstly the intercept of the line on the abscissa defines a dead space which we have called the series dead space and which is similar to that described by Fowler. Secondly the slope of the line indicates the mean alveolar concentration, and thirdly the total volume of nitrogen evolved in the whole breath is measured.

It is now possible to calculate what quantity of nitrogen would be evolved from a particular lung if gas mixing within it were perfect and this calculation, which is commonly used in physiology, is shown in Table 1.

We now have two pieces of information - the volume of nitrogen recovered from the lung by experiment and the volume predicted if gas mixing were perfect. If we divide one by the other and multiply by 100 we have the alveolar mixing efficiency expressed as a percentage.

The values for this efficiency are shown in Figure 5. Here is shown the theoretical line, the values seen in normal subjects and the values seen in disease. The last line on the figure indicates the effect on nitrogen recovery of a regional maldistribution of inspired gas using the data of West in his ten compartment model. It can be seen that distribution of ventilation can account for only a small part of nitrogen retention.

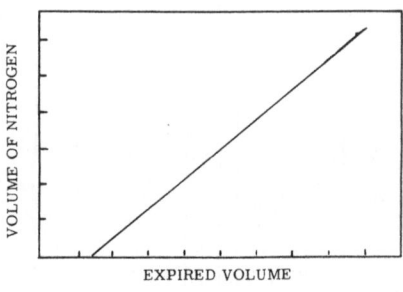

Fig. 2(c)

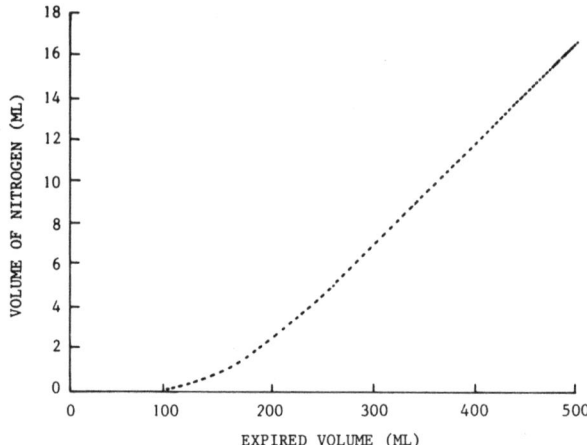

Fig. 3

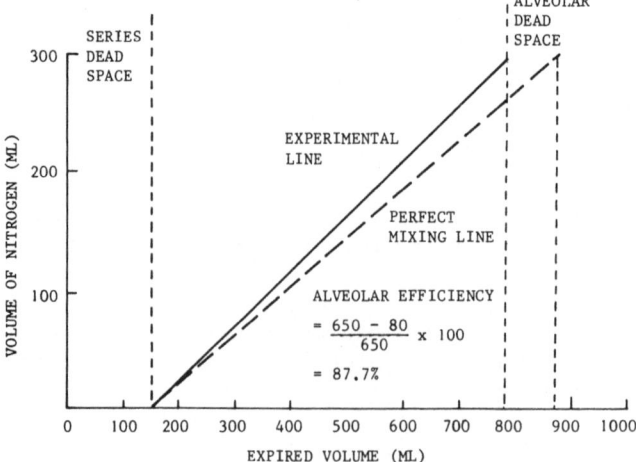

Fig. 4

Table 1. Nitrogen Recovery Following
an Inspirate of Oxygen

The lung volume contains a mixing volume V_A and a non-mixing volume V_{Ds} such that $V_L = V_A + V_{Ds}$. Following the inspirate, V_A increases in volume by V_T, the new volume being $V_L + V_T$.

The volume of oxygen added to this volume is $V_T - V_{Ds}$, so that the dilution ratio is:

$$= \frac{V_L}{V_L + (V_T - V_{Ds})} \qquad \text{or} \qquad \frac{V_L}{V_A + V_T}$$

If nitrogen concentration is C_0 the new concentration C_1 is

$$C_1 = \frac{V_L}{V_A + V_T} \times C_0$$

During expiration, V_{Ds} will retain its volume of nitrogen and excrete a volume $V_T - V_{Ds}$ of gas at concentration C_1 or

$$V_{N_2} = \frac{V_L}{V_A + V_T} \times C_0 \times V_T - V_{Ds}$$

Inserting the appropriate values.

$$V_{N_2} = \frac{3805}{4258} \times 80 \times 453$$

$$= 324 \text{ ml.}$$

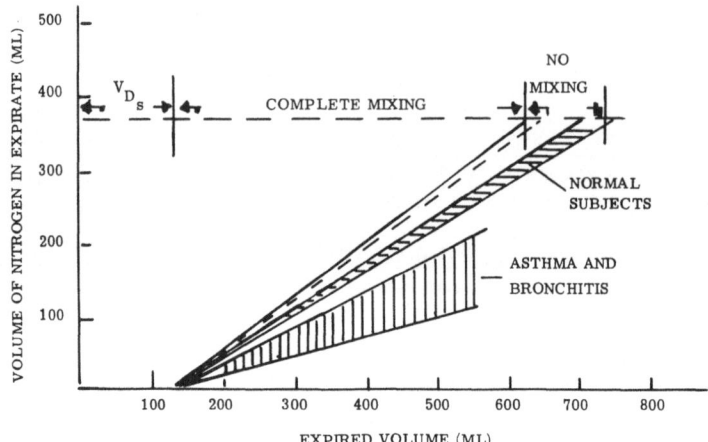

Fig. 5. Distribution of ventilation and perfusion in an alveolar
         duct.

In normals such distribution cannot explain mixing defects completely, and in disease it is even more difficult to use the parallel
hypothesis to explain an efficiency of perhaps 25%.

Before embarking on an alternative explanation it is necessary
to make some further remarks about the nature and the meaning of a
series dead space.

In the test previously described, pure oxygen flows down the
bronchial tree, its linear velocity becomes progressively less as the
area of cross section of the bronchial tree increases. In the
trachea during quiet inspiration the velocity is about 40 cms per
second, falling in the terminal bronchiole to about 1 cm/sec and
beyond this falling further as branching in the terminal ventilatory
unit continues. At the same time as convective flow bears oxygen
molecules forward, and with them the nitrogen molecules with which
they are in contact, the nitrogen molecules are moving upwards towards the mouth due to gaseous diffusion across the concentration
gradient. This upwards movement may be regarded as velocity.

We now have two phenomena occurring at the same time - a velocity of nitrogen upwards which is fixed and determined by the diffusion characteristics of the bases and a velocity downwards which is
progressively diminishing brought about by bulk flow.

It is clear that at some point within the terminal ventilatory
unit these two velocities will become equal and the bulk flow of nitrogen downwards will be exactly balanced by the diffusive flow upwards. At this anatomical site the interface between oxygen and

nitrogen becomes stationary and remains so whilst inspiratory flow
continues at the same rate. Beyond the stationary interface therefore, oxygen molecules gain access to the ventilatory unit only by
diffusion and the rules governing gas mixing are the rules of gaseous
diffusion.

Before these rules are discussed, it is important to understand
the role played by bulk flow in the process of ventilation. This can
best be done by considering the situation of the lung which is full
of nitrogen and into which nitrogen is inspired. Since there is no
concentration difference between the inspired and resident gas there
is no mixing at all, yet alveoli expand and inspire gas, which is entirely nitrogen. This volume represents the ventilation of the lung
by bulk flow.

When the inspired gas is oxygen and diffusion plays its role,
bulk flow into alveoli occurs in exactly the same way as in the case
of nitrogen. The oxygen however gains access only after diffusion
and if the bulk flow is large, the stationary interface moves dis-
tally and ventilation in that unit is larger but still determined by
the balance between diffusive forces and bulk flow forces.

Let us now enquire about the rules which govern diffusive
mixing, since events distal to the stationary interface dominated by
this phenomenon.

Three main variables are involved in diffusive mixing - (1) the
diffusion coefficient between the two gases, (2) the distance over
which diffusion takes place, or the diffusion path length, and (3)
the area of contact between the two gases - the interface area.

In normal breathing we are not concerned with the diffusion coefficient since it remains the same in health and disease so there
remains the two determinants of diffusion path length and interface
area. Both are determined by the geometry of the lung by the nature
of the inspiratory manoeuvre. When these two variables have been
carefully studied using mathematical models the conclusion reached by
nearly all authors is that equilibrium is almost complete within a
few seconds and significant nitrogen retention could not be explained
by a failure of diffusive mixing. Nevertheless experiment shows that
it does occur and we now see that neither the regional hypothesis nor
the diffusion hypothesis appears to be capable of explaining the
common observation.

When one arrives at such a position in which all known hypothe-
ses appear to have been falsified it is useful to re-examine the
original assumption upon which the disproof rests.

In the mathematical analysis of diffusion the anatomical information used has been that of a symmetrically branching structure with

Table 2. The Structure of the Terminal
         Ventilatory Unit

A. MORPHOMETRIC DATA

Number of terminal bronchioles	26,000
Number of third order bronchioles on each duct	8
Number of RB3 in the lung	206,000
Number of alveoli in the lung	300 million
Number of alveoli on each RB3	1,400
Number of alveoli per duct	18
Number of ducts on each RB3	80

---

six generations of branching beyond the terminal bronchiole. It is now possible to look at the known anatomical information and to test the validity of the assumptions.

Table 2 shows some generally agreed values for the branching of the airways and from which we can conclude that on each respiratory bronchiole of the third order there are about 80 alveolar ducts. The symmetrical tree generation model calls for 14.

Table 3. The Structure of the Terminal
         Ventilation Unit

B. BRANCHING PATTERNS

Mean number of branchings after RB3	6
Minimum number	2
Maximum number	11
Total number of ducts on RB3	80

Table 3 shows the agreed non-symmetrical structure distal to an RB3, a mean number of orders of six, a minimum of two and a maximum of eleven. If we assume that the structure is symmetrical and there are eleven branches then the number of ducts is 2047. Thus we see that assumptions about anatomy, particularly with regard to its symmetry or otherwise, can make an enormous difference to the assumed geometry between 14 and 2000 ducts.

What is required therefore is a structure for the terminal ventilatory unit which satisfies the known anatomical facts and then to draw some conclusion about its function. There are an infinite number of equivalent structures but the main features are shown in Figure 6. This structure has the right number of ducts and the correct degree of asymmetry and the first functional question to be asked is what is the position of the stationary interface.

Fortunately it is possible to measure this in terms of volume since during expiration the stationary interface is expelled at the lips and produces the rapid upstroke of the expired concentration curve. The volume of the series dead space therefore represents the position of the stationary interface within the lung volume. What this means in anatomical terms is rather more difficult but on average it will be formed towards the end of the third order respiratory bronchiole, in other words at the entrance of the structure depicted on this slide.

We have seen that the ventilation of the unit will be determined by bulk flow, but that mixing within it will be determined by diffusion path length and interface area. Since all pathways have similar diameters, the major determinant of mixing efficiency will be diffusion path length - in other words the distance between the stationary interface and the anatomical point considered.

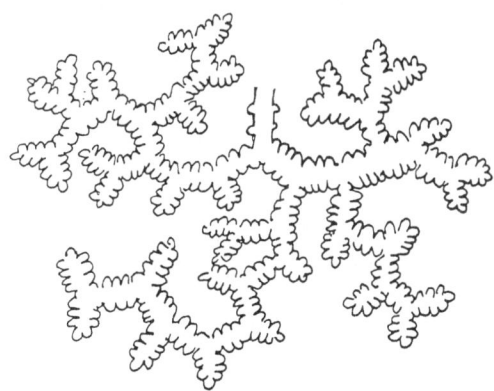

Fig. 6. Number of ducts = 70. Duct structure distal to a respiratory bronchiole of third order.

Inspection of this structure will convince you that there is a great variety from 2 mm for the shortest to 14 mm for the longest. Consequently there will be within the terminal unit a variety of times of equilibration, and in all probability a failure of complete mixing along the longer pathways.

As we have seen the alveolar efficiency of the normal lung is high, over 85% and this small inefficiency might well be explained on this anatomical basis.

How does this hypothesis work when it is applied to disease? Those diseases which manifest mixing defect include asthma, bronchitis and emphysema - so that all possess the same functional defect of impaired mixing, in addition to sharing in common airflow limitation and dyspnoea.

To function effectively in explaining disease the anatomical hypothesis must show how increased diffusion path length and interface area are produced by the disease process.

Taking first the case of bronchial asthma in which the functional defects of airflow limitation and gas mixing deficit are combined, and the severity of each does not correlate with the other. The extent of mixing deficit is often gross, and the size of this deficit is difficult to explain on the basis of regional distribution as produced by the large airways, although it is in these airways that flow limitation has its origin.

We must turn to another aspect of asthma for the necessary clue. In asthmatic patients who are killed in accidents and who come to autopsy it is common to find small airways of about 1 mm diameter completely occluded by solid mucous secretion and such blockage is diffuse and a good proportion of airway is involved.

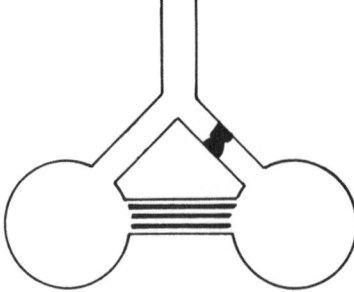

Fig. 7. Diagrammatical representation of collateral ventilation. Communications are at alveolar level.

Taking this evidence together with the mucous spirals which are coughed up towards the end of an asthmatic attack it seems likely that blockage of small airways is a common feature in asthma.

Ventilation distal to such blocked airways is by collateral channels, as represented in Figure 7. In such a system the stationary interface is formed in the patient airway, whilst ventilation of the distal airway is (a) along a much longer diffusion pathway and (b) has a very much smaller interface area, and these conditions are as we have seen, those which produce an impairment of gas mixing.

Thus the gas mixing efficiency in bronchial asthma becomes a measurement of the quantity of lung in which there are occluded small airways. It is a natural corollary of this statement that treatment such as steroids which is effective in opening up closed airways, can be monitored by measuring gas mixing efficiency.

Figure 8 illustrates the situation seen in centrilobular emphysema where a small dilatation occurs at the level of the respiratory bronchiole and in ventilarory terms this is identical with bronchial asthma and the mixing defect seen in this disease is explained in the same way.

Pan acinar emphysema, in which there is complete destruction of the architecture of the terminal unit, and which also has a mixing defect is probably explained by the long diffusive path length produced by the dilated sac, which may be several centimetres in length.

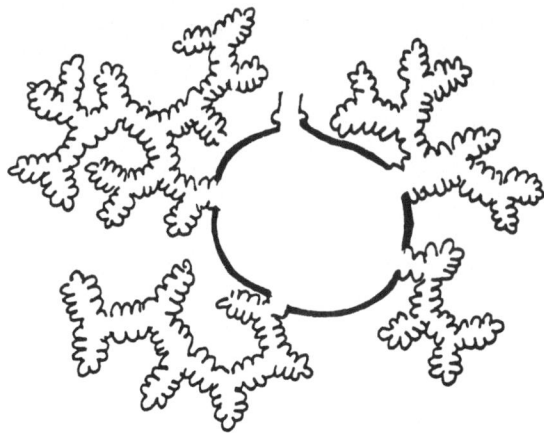

Fig. 8. Centrilobular emphysema. Two orders of alveolar ducts have been destroyed.

Finally we come to the common disease of chronic bronchitis, in which cigarette smoking plays such a large part in aetiology. Here the situation is again of small airway occlusion but there is in addition to mucous plugging cicatrisation of small airways so that their occlusion is irreversible. The ratio between mucous plugging and cicatrisation is not currently known.

When the ventilatory efficiency of cigarette smokers who complain of no symptoms is measured a proportion have impaired efficiences. Are we to conclude that there is already an attack upon their small airways, and have we identified a group which will eventually succumb to their habit? The therapeutic attack upon airways disease will become quite different if it is recognised that large airway disease and small airway disease usually co-exist and that the physiological manifestations of the two components are independent each of the other.

Whilst we all tackle large airway disease enthusiastically we give no attention at all to small airway disease, and the time has now come when this defect should be remedied.

COMPUTING CONTINUOUS DISTRIBUTIONS FROM LIMITED DATA

Peter D. Wagner

Department of Medicine
University of California San Diego
La Jolla, California 92093

INTRODUCTION

The hallmark of many lung diseases is maldistribution. For example, in chronic obstructive pulmonary disease (COPD), ventilation ($\dot{V}A$) is distributed non-uniformly with respect to volume (V), resulting in a multiexponential N_2 washout. In COPD, asthma, interstitial fibrosis and many other types of lung disease, ventilation ($\dot{V}A$) is distributed non-uniformly with respect to blood flow ($\dot{Q}$), resulting in impaired steady-state gas exchange, generally manifest by hypoxemia $\pm$ hypercapnia. Methods for quantitating the degree of $\dot{V}A/V$ or $\dot{V}A/\dot{Q}$ inequality in such patients have been proposed almost continuously since the numerical aspect of gas exchange was first brought to prominence some 30 years ago.

Unfortunately, the problem of exactly describing the distribution of $\dot{V}A/V$ or $\dot{V}A/\dot{Q}$ is insoluble. Individual gas exchange units are of course not amenable to having their ventilation or blood flow measured directly. Thus, the distribution of $\dot{V}A/V$ or $\dot{V}A/\dot{Q}$ can only be assessed indirectly, as by analysis of the N2 wash-out or arterial PO_2 respectively. Finally, because of the immense number of gas exchange units contained within the lung, it is not possible to collect enough data to make these indirect methods exact. Thus distributions of $\dot{V}A/V$ or $\dot{V}A/\dot{Q}$ can only be estimated to a hundred per cent.

METHODS

There are two different approaches to estimating these distributions. Traditionally, a simple two or three-compartment model has been used and sufficient data collected to determine the parameters

15

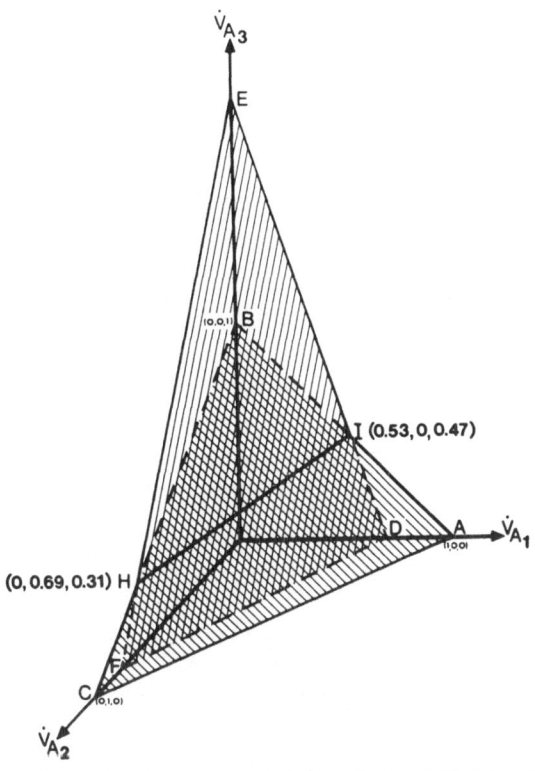

Fig. 1. Analysis of the multibreath nitrogen wash-out by linear pro-
         gramming. For a 3-compartment lung with fractional ventila-
         tions $\dot{V}A_1$, $\dot{V}A_2$ and $\dot{V}A_3$, a 3-dimensional positive octant is
         drawn. On it appear two planes. One expresses the fact
         that the sum of fractional ventilations are unity and is
         thus the plane $\dot{V}A_1 + \dot{V}A_2 + \dot{V}A_3 = 1.0$. The second assumes
         that the lung, containing just 3 lung units of $\dot{V}A/V$ ratios
         of 0.02, 0.2 and 2.0 respectively, has in this example, a
         mixed expired N_2 concentration of 1.0 units prior to N_2
         wash-out and 0.6771 after the first breath of O_2. This
         second plane is a mixing equation for the 3 compartments and
         is given by

$$0.98\ \dot{V}A_1 + 0.83\ \dot{V}A_2 = 0.33\ \dot{V}A_3 = 0.6771$$

The line of intersection between the two planes (H-I) gives
all possible combinations of $\dot{V}A_1$, $\dot{V}A_2$ and $\dot{V}A_3$ which simultaneously satisfy the above two equations. The number of
such solutions is infinite, but the endpoints of H-I give
maximal possible values of $\dot{V}A_1$, $\dot{V}A_2$ and $\dot{V}A_3$ compatible with
the equations. In linear programming, all such endpoints
are systematically examined to reveal upper bounds (i.e.
maximum values). Although illustrated for 2 equations and
3 compartments, the principle holds for any number of equations and any larger number of compartments.

of the model compatible with the data. The advantage of such methods
is their relative simplicity and clinical usefulness. However, there
are several disadvantages. For example, the two or three-compartment
model is too simple to approach reality. Moreover, although designed
to estimate maldistribution, the model parameters are often sensitive
to factors other than the degree of maldistribution per se. This is
the case for the Riley model[1] of deadspace and venous admixture,
whose values change with changes in total ventilation or blood flow.[2]
Finally, in many of these methods, no account is taken of experimental error nor of other compatible models.

For these reasons, we have pursued the multicompartment approach
to estimating distributions. By dividing the lung into some 50 lung
units, the lung is treated in effect as a continuous distribution capable of possessing units of any $\dot{V}A/V$ (or $\dot{V}A/\dot{Q}$) between 0 and infi-
nity. While this approach does not overcome the basic problem of
describing a very large population of units, as stated above, it is
amenable to analyses which take account of both experimental error
and multiple compatible models. We have developed a 50-compartment
least squares analysis for the N_2 washout[3] and for the elimination of
a mixture of infused dissolved inert gases. Using well-known regularization procedures, the least squares criteria are applied in fitting data to the 50-compartment model. Regularization converts the
least squares analysis into the simple solution of a set of simultaneous linear equations. The number of equations and unknowns are
always equal to one another and in turn equal the number of data
points available. Thus, a 50-compartment analysis of a 20 breath N_2
wash-out becomes a problem in solving a 20 x 20 set of simultaneous
equations; the corresponding formulation for the inert gas method
(6 gases) leads to a set of only 6 linear equations. Enforcing po-
sitivity constraints on compartmental $\dot{V}A$, $\dot{Q}$ or V are of utmost importance. If this is not incorporated, the result is generally of
no physiologic value.

There are several advantages of this. First, the approximate
shape and position of the distribution can be estimated. In the case
of the inert gas method, comparison of O_2 and inert gas exchange
permit reasonably direct estimates of the role of diffusion impairment in hypoxemia. In addition, separation of intrapulmonary ($\dot{V}A/\dot{Q}$
mismatching, shunt, diffusion impairment) from extrapulmonary (low
mixed venous PO_2 for any of several reasons) causes of hypoxemia can
be made. This is of major importance in ICU where complex changes
in both intra- and extrapulmonary factors often occur simultaneously.
Arterial PO_2 only reflects the net effects of all these factors.

The disadvantage, more apparent than real today because of the
ready availability of computers, is that the multicompartment least
squares analysis is more complex than the two-compartment approaches
and requires digital computation. This is generally accomplished

easily on microcomputers such as the DEC PDP-1103, with an execution time of just a few seconds.

No matter which approach is favoured in attempts to quantitate maldistribution, the investigator must acknowledge the impossibility of an exact solution, for reasons discussed above. This immediately raises the question of how accurately any of the above-mentioned reflects the exact distribution. Neither the two nor the 50 compartment approach answers this question - they merely give one estimate of the distribution compatible with the data.

We have proposed that Linear programming methods are well-suited to answering the question of just how much information is available from such method.[4,5,6]   It is a straightforward matter to use linear programming to determine the upper bounds (or envelope) within which all compatible distributions must lie. Inspection of these upper bounds and other formulations of the linear programme give considerable insight into the real nature of the distribution and form the basis for interpretation of the two or 50-compartment methods described above.

CONCLUSIONS

We believe that the multicompartment analysis with regularization ("smoothing") is the most informative, although by no means the only method for determining maldistribution within the lung. This conclusion is based upon rigorous upper bound analysis by linear programming which reveals the real information content of the method. Both linear programming and least squares regularization require digital computation, but small minicomputers such as the PDP-1103 are entirely adequate for the purpose.

REFERENCES

1. R. L. Riley and A. Cournand, Analysis of factors affecting partial pressures of oxygen and carbon dioxide in gas and blood of lungs, Theory J. Appl. Physiol. 4:77 (1951)
2. J. B. West, Ventilation-perfusion inequality and overall gas exchange in computer models of the lung, Respir. Physiol. 7: 88-110 (1969).
3. S. M. Lewis, J. W. Evans, and A. A. Jalowayski, Continuous distribution of specific ventilation recovered from inert gas wash-out, J. Appl. Physiol., Respirat. Environ. Exercise Physiol. 44:416-423 (1978).
4. J. W. Evans and P. D. Wagner, Limits on $\dot{V}A/\dot{Q}$ distributions from analysis of experimental inert gas elimination. J. Appl. Physiol., Respirat. Environ. Exercise Physiol. 42:889-898 (1977).

5. P. D. Wagner, A general approach to the evaluation of ventilation perfusion ratios in normal and abnormal lungs. Physiologist 20:18-25 (1977).
6. P. D. Wagner, Information content of multibreath nitrogen washout. J. Appl Physiol. 46 (3):579-587 (1979).

ACKNOWLEDGEMENTS

This work was supported by NIH Grant HL 17731. It is a pleasure to acknowledge the secretarial assistance of Tania Davisson on the preparation of this manuscript.

COMPARISON OF DIFFERENT METHODS OF MEAN ALVEOLAR PCO$_2$ MEASUREMENT

DURING MECHANICAL VENTILATION: AN EXPERIMENTAL AND THEORETICAL STUDY

G. Boy,* ** M. F. Forda,* B. Renun,* J. M. Brieussel,*
E. Caval* and L. Lareng*

*Dept. d'Anesthesie - Reanimation - Hopital Purpan
31000 Toulouse, France
** Nat. Off. of Aeronautics Studies & Research
(Onera/Cert/Dera)
2, Ave Ed. Berlin, 31055 Toulouse, France

All these methods assume that mean alveolar parital pressures are necessarily localised on the alveolar plateau of the expired gases curve. We explain why these methods are necessarily wrong when the expiratory flow wave is localised at the beginning of expiration. A dynamic mathematical model based on the three phenomena: convection, molecular diffusion and alveolar capillary exchange, has been built and tested by different numerical methods. The solutions of this model point out the great importance of the flow wave during expiration. Particularly its shape determines if the mean alveolar partial pressure value crosses the alveolar plateau or not. This theoretical result is very important to interpret the gradient between alveolar gas calculated at the mouth and arterial blood gas.

INTRODUCTION

Non-invasive methods for measurement of mean alveolar partial pressure of oxygen and carbon dioxide are actually of great interest. In critical care the evaluation of such parameters is known to be a very difficult problem. Since Fenn, Rahn and Otis[16] and Cournand and Riley[17] who have introduced the new concepts of the actual meaning of respiratory function, some authors such as Bargeton,[2] Guenard and Chaussain,[18] Colinet-Lagneau, Troquet et al.,[20] Menier et al.,[19] and more recently Luft, Loeppky and Mostyn[6] have studied the calculation of PAO$_2$ and PACO$_2$ values from the information of expired gas signals.

Bargeton has used the Fenn diagram and respiratory ratio equality assumption. This method was studied and compared to BOHR and end-tidal methods by Luft et al.

The gas sampling method proposed by Guenard and Chaussain is based on the criterion of mean alveolar (RA) and expired (RE) respiratory ratio equality. A bolus of expired gas was sampled at an expired volume equal to $V_T/2 + V_D$ (V_D: volume of the dead space).

We have tested four methods. Two that can be classed with the sample methods: End-tidal measurement and gas sampling at a point ($3/4$ V_T here) localised on the alveolar plateau, and two graphical methods, the planimetry method used by Chopin et al.[3,4] and the procedure described by Bargeton on the (PCO_2, PO_2) diagram of Rahn and Fenn.[1] Generally the expiratory record crosses the RE line. The mean alveolar values for O_2 and CO_2 are the coordinates of the point where the tracing reaches the instantaneous "alveolar" IR (Gbikpi-Benissan[5] notations) value that is equal to RE obtained from the mixed expired air. Theoretically, this method is valid for all the physiological and pathological cases.

Notations:

Indexes
        a   - arterial
        A   - Alveolar
        B   - at the mouth
        C   - capillary
        i   - may be CO_2 or O_2

Coefficients
        Bi   - coefficient taking into account solubility in plasma and chemical reaction with haemoglobin,
        Di   - molecular diffusion coefficient,
        λi - alveolar; capillary gas exchange coefficient,
        L   - length of the model tube,
        q(t) - ventilatory flow ($q>0$ inspiration; $q<0$ expiration)
        S(x) - total cross section of airways,
        T   - respiratory cycle time.

Variables
        Pi   - partial pressure of gas i,
        t   - time,
        x   - distance from the mouth.

After many experiments, it appears that the criterion RA = RE was not always attained before the end of a normal expiration.[6] That is this point that we want to emphasize here.

In fact, a real scientific study of the alveolar gases at the mouth necessitates the construction of a mathematical model that can

take into account the different phenomena of gas transfer in the
lung. Many authors have advanced in this domain. The principal
problem is to determine the relative importance of the two phenomena:

- evidence for stratified inhomogeneity in the lung, and
- analysis of unequal distribution of ventilation perfusion and
  alveolar-capillary gas exchange capacity in the lung.[7,5]

Our model is based on the morphometric data of Weibel.[9] Like
the models of Cumming et al.,[11,12] Paiva,[10] Scherer et al.,[14] Laforce
and Lewis[13] and Chang et al.,[15] the structure of the model can explain only the eventual stratification problem, but for our demon-
stration it is sufficient.

We shall discuss the basis of our model and give the main results of interest. The main objectives of this study is to give a
new interpretation of the "mouth-alveolar" - arterial gradient observed in many patients in critical care.

EXPERIMENTAL METHODS AND RESULTS

A series of 41 patients with various types of pulmonary disease
unselected but in critical care and under artificial ventilation.
All the observations were made in a horizontal position. Arterial
blood gases were measured with a catheter inserted in a brachial
artery.

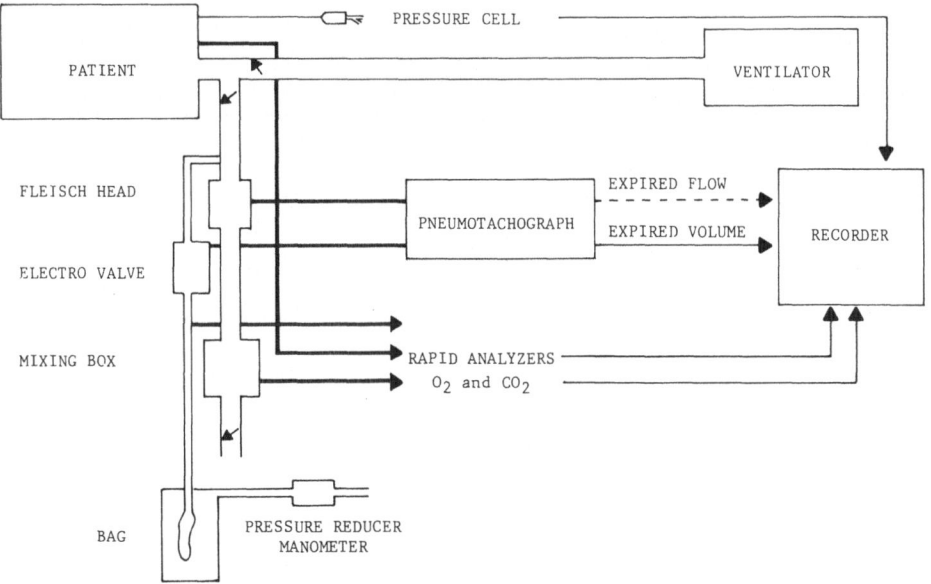

Fig. 1. Diagram of the experimental device.

Table 1. Experimental Results

	MEAN	MEAN VALUE OF DIFFERENCE
P_{AE} Planim.	19.2	
P_{AE} 3/4	20.6	-1.418 ± 3.100
P_{AE} 3/4	20.6	
P_{ET}	24.8	-4.143 ± 1.980
P_{ET}	24.8	
Pa	37.7	-12.970 ± 8.160
$R_{\overline{E}}$	0.9878	
R_{ET}	0.6718	0.316 ± 0.260
R_{ET}	0.6718	
R_{AE} 3/4	0.7759	-0.104 ± 0.330
$R_{\overline{E}}$	0.9878	
R_{AE} 3/4	0.7759	0.211 ± 0.290

Expired flow was measured with a Fleisch head and an electronic pneumotachograph which compute expired volume by integration of flow signal. An electronic device controls an electro-valve for the collection of gas sampled at 3/4 V_T. Oxygen and carbon dioxide fractions were analysed with Beckmann rapid analysers (LB2 - OM 11). Ventilators used are of type Bennett (MA 1B, Bear) Fig. 1.

The main results of this experiment are reported in Table 1. The difference between planimetry method and gas sampling at 3/4 VT is not significant, but in practice the second is more easy. On the contrary, the difference between PA 3/4 and PET is very significant (t = 13,4) but it would be dangerous to conclude that the end-tidal value cannot represent mean alveolar value. Indeed, the comparison of end-tidal partial pressure and arterial partial pressure the difference is also significant (t = 10,2). A question that can be posed is: are pulmonary shunt and air-blood barrier diseases the only reasons to explain this end-tidal - arterial gradient?

Paired comparison of $R_{\overline{E}}$ with R_{ET} showed a significant difference between the mean values (t = 7,9) and the mean of $R_{\overline{E}}$ is greater than the mean of R_{ET}. We found an identical result with the ($R_{\overline{E}}$, $R_{3/4}$) pair.

CONSTRUCTION OF A MATHEMATICAL MODEL

Faced with such results, we have taken a theoretical approach: the modelisation of gas transfer in the lung. The objectives of our model design are two fold: (1) to have a better knowledge of the transfer phenomena in the lung, (2) and to have the possibility of a feedback comparison of experimental results and theoretical predictions.

Following the works of Culling, Paiva and Chang, the lung structure has been considered as a variable section S (x, t) cone. This section represents the sum of all the bronchial sections at a x - distance from the mouth.

The following assumptions have been made:

- conducting airways (first 16 generations) are rigid and the alveolar zone is global and elastic (fig. 2.)

- molecular diffusion is binary and with a constant coefficient,

- the two independent variables are the time t and the spatial coordinate x, along the mouth-alveoli tube,

- dependent variable is the partial pressure of oxygen or carbon dioxide.

- gas transfer of oxygen or carbon dioxide is due to convective and diffusive phenomena and alveolo-capillary gas exchange: in such a way, the transfer system may be represented by the following system of equations:

$$S \frac{dPi}{dt} + q(t)\frac{dPi}{dx} - Di\frac{d}{dx}(S\frac{dPi}{dx}) = 0 \qquad \text{sur }]o,L[x]0,T[$$

with boundary conditions:

At the mouth: Pi (0,t) = PiB   during inspiration

$\frac{dPi}{dx}$ (0,t) = 0   during expiration

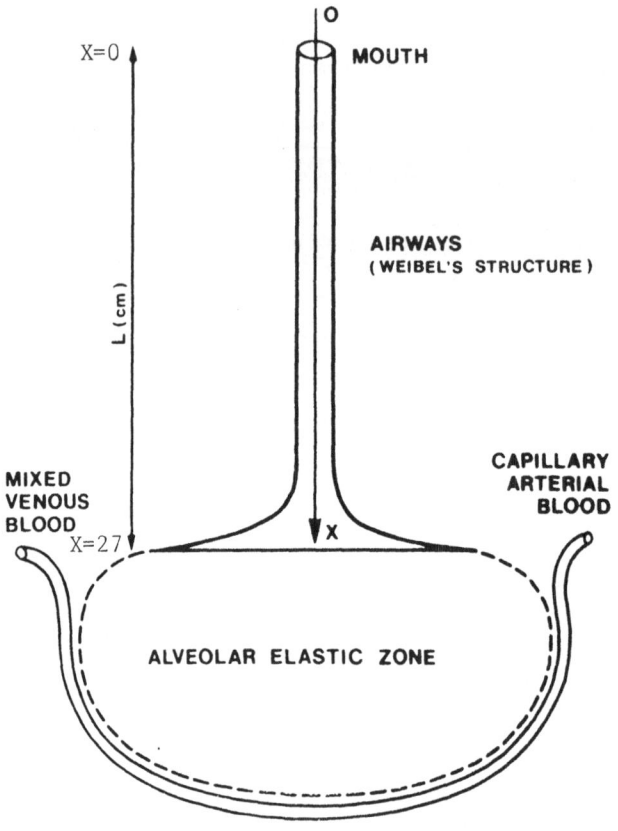

Fig. 2.

In the alveolo-capillary zone:

$$\frac{d(VA\ PiA)}{dt} = q\ (t)\ Pi\ (L,t) - Di\left|S\ (x)\frac{dPi}{dx}\ (x,t)\right|_{x\ =\ L}$$
$$+\lambda i(Pic - Pi_A)$$

$$Vc\ \frac{dPic}{dt} = \dot{Q}\ Bi\ (Piv\ (t) - Pia\ (t)) - \lambda i\ (Pic - Pi_A)$$

and with initial conditions:

$$Pi\ (x,0) = P^o\ (x)\quad sur\ ]o,L]$$
$$Pi_A(0) = P^o i_A$$
$$Pi\dot{C}\ (0) = P^o i_c$$

     This system described by partial differential equations has been
solved by a numerical method based on mixed finite elements. This
method is particularly adapted to simulate stiff fronts like the inspired ones in the lung and give a great fiability of the results.
This aspect shall be developed in later papers.

     The main purpose of this model is to take into account simul-
taneously three phenomena of great importance: convection, molecular
diffusion and alveolar-capillary gas exchange in a geometrical structure based on the more recent descriptions.

DISCUSSION

     The results of different numerical simulations have shown the
great importance of the expiratory flow wave on the expired gas (O_2,
CO_2) curves. Indeed, when the flow presents a rectangular shape,
alveolar gases are transported byconvection with a practically constant rate, from alveolar zone up to the mouth (Fig. 3).   Then the
expired alveolar plateau shape reflects the real alveolar "plateau"
shape of the alveolar gas curves (Fig. 4).   In this case the end-
tidal stratification is negligible and the end-tidal-mean alveolar
carbon dioxide (resp. oxygen) gradient is positive (resp. negative
see Table 2).   On the contrary, when the flow is concentrated in the
beginning of expiration, in such a case "there is not enough flow" to
expire alveolar gases during the last part of expiration time (Fig.
5).   Then when the convection phenomena becomes negligible in front

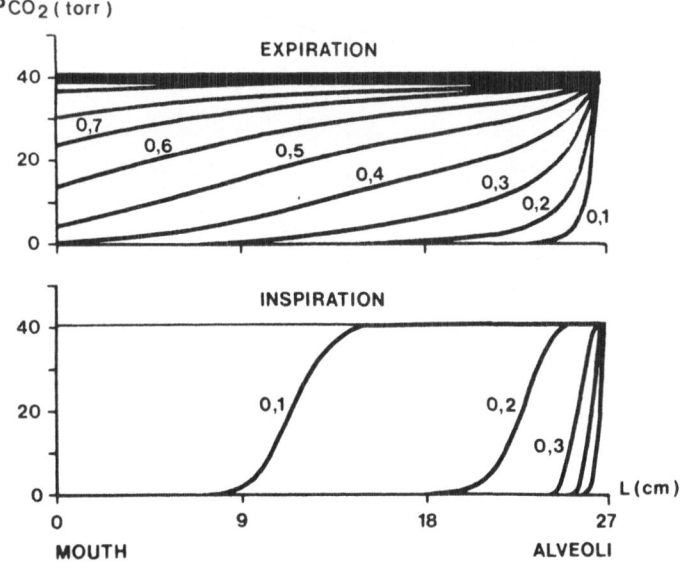

Fig. 3.

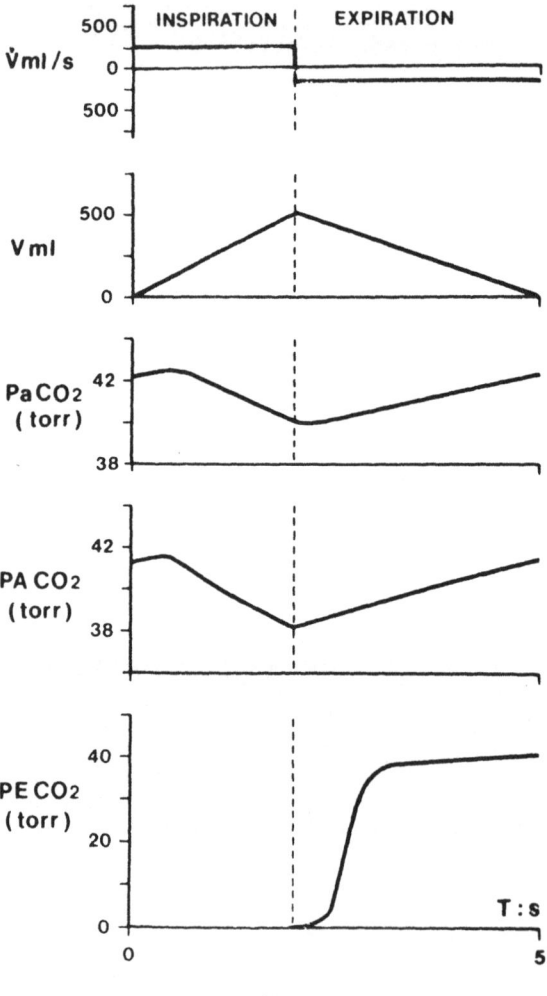

Fig. 4.

Table 2. Model Results

EXPIRATORY FLOW WAVE	CARBON DIOXIDE		OXYGEN	
$(P_{ET} - \bar{P}_A)/P_{ET}$ %	+ 1.5 %	– 3.8 %	– 1.0 %	+ 2.0 %
P_{ET} (torr)	40.7	39.0	97.0	98.0
$\bar{P}_A$ (torr)	40.1	40.5	98.0	96.0

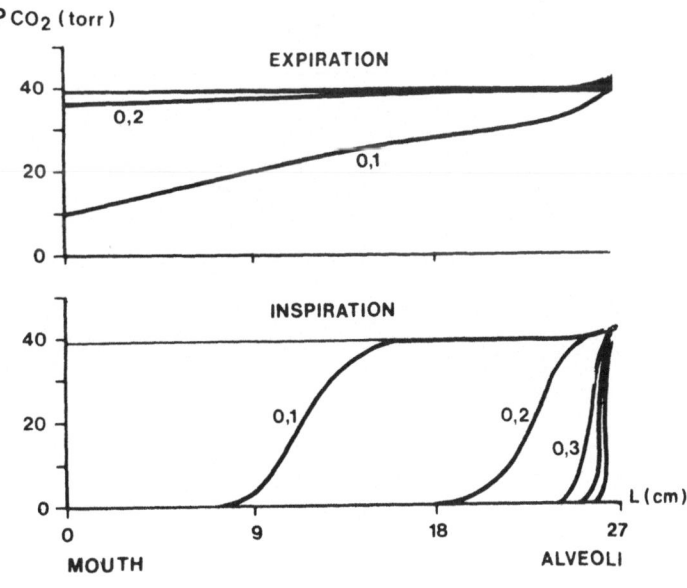

Fig. 5.

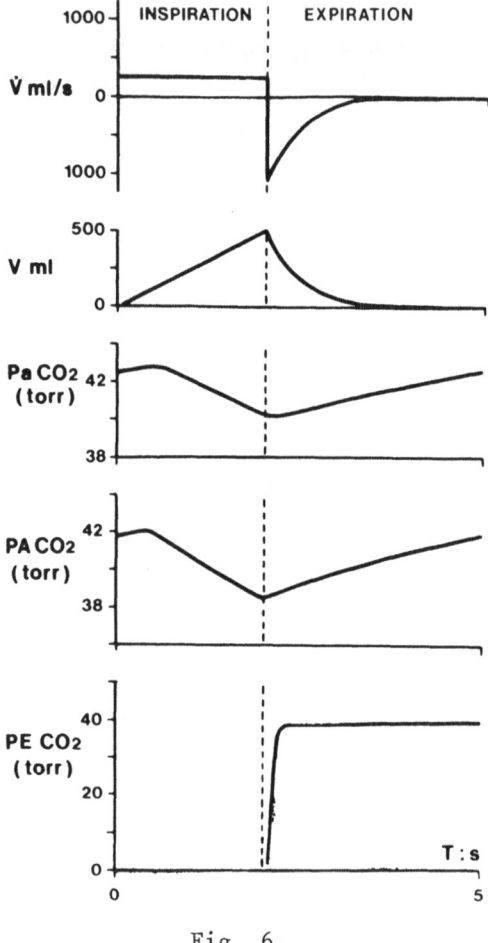

Fig. 6.

of molecular diffusion and alveolar-capillary gas exchange, a significant stratification appears. It is obvious that in this way the
alveolar carbon dioxide (resp. oxygen) continues to climb (resp. to
descend) whereas the expired alveolar plateau stays practically horizontal (Fig. 7), then, it is possible to think that the mean alveolar
carbon dioxide partial pressure may be over the end-tidal mean alveolar carbon dioxide (resp. oxygen) gradient is negative (resp. positive) (see Table 2). It is a conformation of the results of Luft et
al.[6] to which we give a new interpretation.

     This result is of great importance in practice. Indeed, many
(and practically all) expiratory flow waves of artificially ventilated patients present an exponential shape.

     This result is particularly appreciable in the case of obstruc-
tive patients.

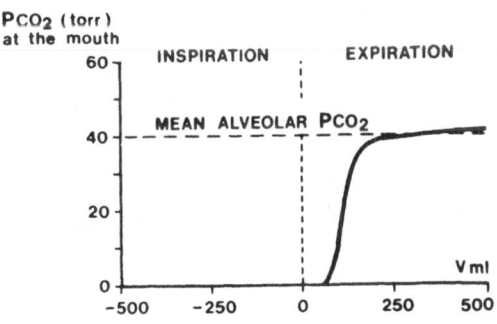

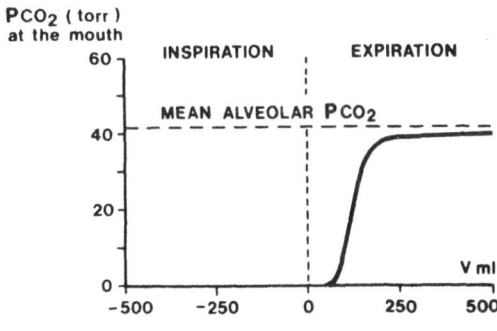

Fig. 7.

Indeed obstructions weaken the flow and an exponential flow
shape observed at the mouth is in fact very much pointed in the alveolar zone; then the phenomenon that we have described is enlarged.
We have not discussed regional inhomogeneity, but the same reasoning
can be applied to for all parts of the lung.   In fact the stratification exists in each regional zone with different degrees due to the
regional flow wave.

REFERENCES

1. H. Rahn and W. O. Fenn.   VA graphical analysis of the Respiratory
    Gas Exchange.   The O_2 and CO_2 Diagram.   Washington D.C. <u>Am.</u>
    <u>Physiol. Soc.</u> 1955.
2. D. Bargeton, Analysis of the capnogram and oxygram in man, <u>Bull.</u>
    <u>Physio Pathol. Respir.</u>: 3:503-526, 1967.
3. C. Chopin, M. C. Cambrin, H. Robin, C. Boulengez, B. Duquesne,
    D. Vallet, A. Durocher, B. Gosselin, F. Wattel, Determination
    sous ventilation assistée de la ductance globale de CO_2 et de
    ses composants partièlles, <u>Am. Anesth.</u> France 1977, XVIII, 7
    & 8, 593-603.
4. C. Chopin, M. C. Chambrin, F. Wattel, Determination sous Ventila-
    tion Assistées et en Pathologic Respiratoire aigue de Para-
    metres Fonctionnels d'effecacité rentilatoire, <u>Acta Tuberc.</u>
    <u>Pneumol, Belg.</u> 1977, 6814:379-395.
5. G. Gikpi-Benissan, J. G. Sageaux, C. Garinski, P. Varenne & J. P.
    Cardinaud, Mesure contisule des pressions particles alveo-
    loires de gaz carborinique et d'oxygène.   Interêt pour la sur-
    veillance des échanges gazeux pulmonaires en réanimation Ag-
    ressologie, 1980, 21 B:75-80.
6. U. C. Luft, J. A. Loeppky and E. M. Mostyn, Mean alveolar gases
    and alveolar-arterial gradients in pulmonary patients, <u>J. Appl</u>
    <u>Physiol. Respirat. Environ Exercise. Physiol.</u>, 46 (3):534-540,
    1979.
7. J. Pijper and P. Scheid, Respiration: Alveolar gas exchange, <u>Ann.</u>
    <u>Rev. Physiol.</u> 1971, T32:131-154.
8. R. E. Forster, and E. D. Crandall, Pulmonary gas exchange, <u>Ann.</u>
    <u>Rev. Physiol.</u>, 1976, T 38:69-93.
9. E. R. Weibel, "Morphometry of the human lung," Stinger-Berk, 196
10. M. Paiva, "Contribution à la biophysique de system respiratoire,"
    These Iniv. Libre Bruxelle 1973.
11. G. Cumming, K. Horsefield, J. G. Jones and D. C. F. Muir, The
    influence of gaseous diffusion on the alveolar plateau at
    different lung volumes, <u>Resp. Physiol.</u> 2, 386-398, 1967.
12. G. Cumming, K. Horsefield and S. B. Preston, Diffusion equili-
    brium in the lung examined by modal analysis, <u>Resp. Physiol.</u>
    12:329-345, 1971.
13. R. C. La Force and B. M. Lewis, Diffusional transport in the
    human lung, <u>J. Appl. Physiol.</u>, Vol. 28, 3, 1970.

14. P. W. Scherer, L. H. Schendalman and A. M. Greene, Simultaneous diffusion and convection in single breath lung wash-out, <u>Bull</u>. <u>Math</u>. <u>Biophys</u>. 34:393-412, 1972.

15. D. B. Chang and S. M. Lewis, A theoretical discussion of diffusion and connection in the lung, <u>Math</u>. <u>Biosc</u>. 29:331-349, 1976

16. W. O. Fenn, R. Rahn and A. B. Otis, A theoretical study of the composition of the alveolar air at altitude, <u>Am</u>. <u>J</u>. <u>Physiol</u>, 146:637-653, 1946.

17. R. L. Riley and A. Cournand, Analysis of factors affecting partial pressures of oxygen and carbon dioxide in gas and blood of lungs, Theory <u>J</u>. <u>Appl Physiol</u>. 4:77-101, 1951.

18. H. Guenard and H. Chaussain, A sampling method for mean alveolar gas in normal subjects and patients with respiratory disease, <u>Bull</u>. <u>Europ</u>. <u>Physiopath</u>. <u>Respir</u>. 12:625-636, 1976.

19. R. Menier, F. Blanchet, Ph. Darcet, D. Desfonds and E. Florentin Analyse Continue des parametres ventilatoires au cours de 1' exercice musculaire, <u>Med</u>. <u>Sport</u>. T50, 1976, 3.

20. J. Troquet, D. Colinet-Lagneaux and J. Hailleux, De la peute affectant le plateau alvéolaire des gaz expirés, <u>Arch</u>. <u>Int</u>. <u>Physiol</u>. <u>Biochim</u>. 1972, 80:835-837.

21. G. Boy, "Structure et Fonction Respiratoires," Thesis - Toulouse 1980.

MULTICOMPARTMENTAL ANALYSIS OF PULMONARY FUNCTION USING SYSTEM

IDENTIFICATION TECHNIQUES

Sven Olofsson and Göran Hedensterna

Department of Clinical Physiology
Huddinge University Hospital
S-141 86 Huddinge, Sweden

Multicompartmental analysis of pulmonary gas exchange is of great interest since it offers a more realistic description of the exchange of gas than the analysis of the lung as a single homogeneous unit.[1,2] It might be expected that the multicompartimental analysis of other lung function variables, e.g. those of mechanical properties, offers similar advantages. The present study was undertaken to develop a non-invasive, bedside method for multicompartmental analysis of lung mechanics, ventilation and blood flow.

The approach was that of systems indentification which can be described accordingly.[3] Consider a lung being fed by a certain input signal, e.g. gas flow. It responds with an output signal, e.g. pressure. A lung model can be constructed, incorporating parameters such as compliance, resistance lung volume, etc., described by algorithms which link the lung function parameter to certain input and output signals. If this lung model is fed by an input signal identical to the one feeding the lung (e.g. flow), the parameters in the lung model are assigned values by an iterative optimigation procedure so that the model responds with an output signal as similar to the one recorded in the lung as possible (e.g. pressure).

The procedure used in the present study involved the recording of the input signals tracheal gas flow and the inspired nitrogen and carbon dioxide concentrations while corresponding output signals were tracheal pressure and expired nitrogen and carbon dioxide concentrations. These signals were recorded during a step change in the inspired N_2 and CO_2 concentrations so that a partial N_2 wash-out and CO_2 wash-in was produced. The signals were sampled: at a frequency of 40 Hz for 4-5 breaths, i.e. during a period shorter than that for recirculation of blood.

   The compartmentation has so far been based on time constants.
From these are calculated the resistance and compliance of each compartment. In succeeding calculations compartmental values are assigned for dead space, alveolar volume, ventilation, alveolar CO_2
tension and blood flow. So far the analysis has been restricted to
a two compartment model. However, it is theoretically possible to
use more compartments in the analysis.

   A limited number of measurements has been done in supine middle-
aged lung healthy men during inhalation anaesthesia. A typical
result is shown in Fig. 1. As can be seen, one compartment is assigned a 10 times higher resistance than for the other compartment.
The dead space ventilation is also unevenly distributed with almost
all delivered to the low-resistive compartment. Slightly uneven distributions of compliance, ventilation, alveolar volume and blood flow
can also be seen. This report does not pretend to explain these data
but ongoing work is directed towards airway closure and collateral
ventilation as possible causes to the second compartment.

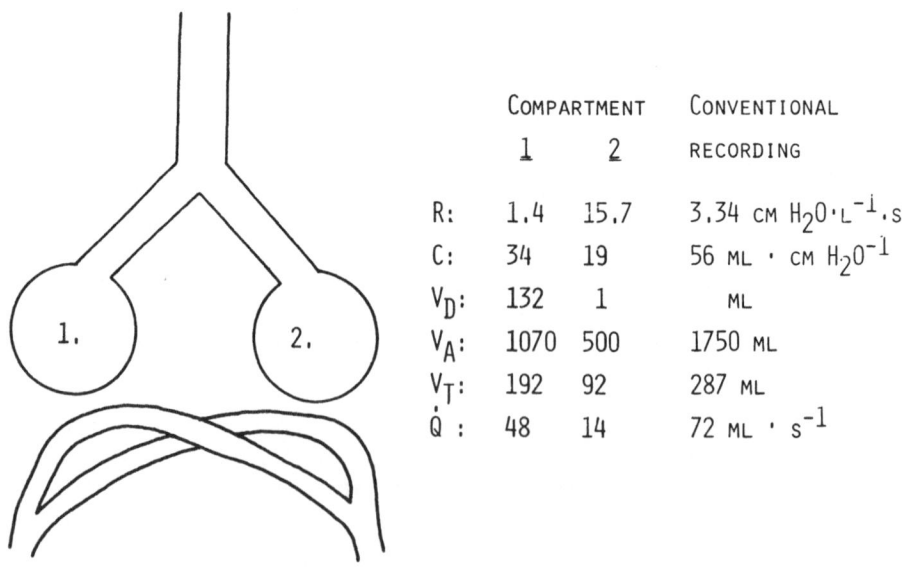

	COMPARTMENT		CONVENTIONAL
	1	2	RECORDING
R:	1.4	15.7	3.34 CM $H_2O \cdot L^{-1} \cdot s$
C:	34	19	56 ML $\cdot$ CM H_2O^{-1}
V_D:	132	1	ML
V_A:	1070	500	1750 ML
V_T:	192	92	287 ML
$\dot{Q}$:	48	14	72 ML $\cdot$ s^{-1}

Fig.1. Multicompartmental analysis of pulmonary function

REFERENCES

1. C. Lenfant, and T. Okubo, Distribution function of pulmonary
    blood flow and ventilation-perfusion ratios in man, J. Appl.
    Physiol. 24:668-677, 1968.
2. P. Wagner, H. A. Saltzman and J. B. West, Measurement of con-
    tinuous distributions of ventilation-perfusion ratios,   Theory
    J. Appl Physiol. 36:588-599, 1974.
3. S. Olofsson, "Measurement of lung function parameters by system
    identification methods. The lung model." To be published.

CHAPTER 2

REGIONAL LUNG FUNCTION AND IMAGING

TOMOGRAPHIC IMAGING OF REGIONAL LUNG DENSITY BY 90° COMPTON

SCATTERING

*Massimo Pistolesi, Massimo Miniati, Stefano Solfanelli
and Carlo Giuntini
**Luciano Azzarelli, Massimo Chimenti, Franco Denoth
and Fabrizio Fabbrini

*CNR Institute of Clinical Physiology and 2nd
Medical Clinic
University of Pisa, Pisa, Italy
**CNR Institute for Elaboration of Information
Pisa, Italy

A tomographic method that permits the visualisation of frontal and sagittal planes of the chest according to their density has been developed in order to detect regional lung density changes in disease.[1,2,3] Sectional visualisation of the chest is obtained by employing a collimated linear source of gamma photons (^{203}Hg or ^{192}Ir) and a gamma camera, as imaging device, to detect the Compton scattering at 90° to the primary beam. Imaging 90° scattered rays provides a view of the anatomical cross section corresponding to the path of the primary rays through chest tissues. Chest tomographic views with 500 K counts are obtained in 1 to 2 min at the spatial resolution of the gamma camera with tissue density discrimination of 12% on the average, as demonstrated by phantom studies. The radiation dose to the patient is about 0.09 rads for each view as a maximum estimate. This radiation load is comparable with that of standard chest x-ray and considerably less than that of conventional or computerised x-ray tomography.

We performed 79 studies in 67 patients with various lung disorders. When the lung density was either reduced such as in emphysema or increased such as in pulmonary edema, the technique visualised density changes over the entire lung. In patients with focal lesions such as bullous emphysema, lung tumor, pneumonia, interstitial fibrosis and localised effusion, the technique provided images comparable to those of conventional x-ray tomography but with better visualisa-

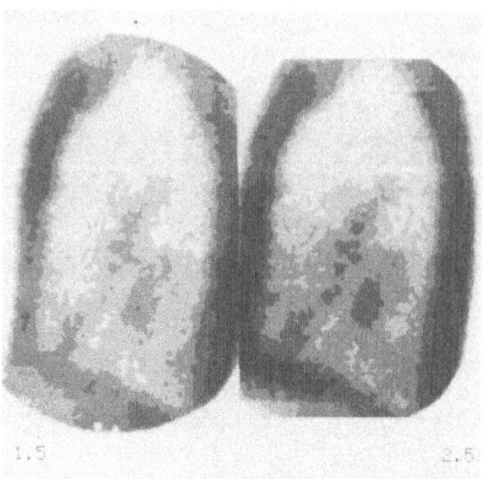

Fig. 1. Tomographic view obtained in a patient affected by emphysema
         of upper regions of the left lung. On the left the image
         is not corrected for primary and scattered ray attenuation.
         On the right the image after computer correction.

tion of the lung structures surrounding the lesion. This may be of
interest for patients who undergo surgery. The low radiation dose
permits application in children and monitoring of lung edema in
cardiac patients.

     The physical attenuation of the primary beam will make the image
of the anatomical structures fainter as distance travelled by the
beam in the chest increases. Furthermore the scattered rays are at-
tenuated by the chest tissues interposed between the tomographic
plane and the gamma camera. In order to overcome this problem, gamma
camera imaging signals recorded on 35 mm film are dignitised by a
computer controlled flying spot system. Digital low-pass filtering
is performed on input in order to reduce photographic and gamma
camera noise; several processing steps are then executed. Firstly,
a matrix normalisation is performed in order to reduce acquisition
errors due to the non-homogeneity of gamma camera detector head and
to the presence of spurious signals; secondly an iterative image re-
storation technique is used in order to eliminate density variations
due to the attenuation of both primary and scattered gamma rays
travelling in chest tissues. Restored images (Fig. 1.) are finally
viewed on TVC monitor or grey level digital plotter: absolute values
of tissue density and relative values of counting density are ob-
trained. Either automatic or interactive procedures for single image
feature extraction or for mass storing and retrieving purposes can be
used for perspective views or isophote representations by extraction
from digitised images. Preliminary application of this computer

correction in patients with pulmonary bullous emphysema demonstrates
the possibility of improving the diagnostic capabilities of the technique. Further development of the computer algorithms will permit
the reconstruction of transverse tomographic planes from the information obtained by the sagittal and frontal planes.

REFERENCES

1. M. Pistolesi, R. Guzzardi, S. Solfanelli, M. Mey and C. Giuntini,
   Regional lung density imaging by 90° scattering of an external
   gamma-ray source, in "Proceedings of the San Diego Biomedical
   Symposium," J. I. Martin, ed., Academic Press inc., New York
   (1977).
2. M. Pistolesi, S. Solfanelli, R. Guzzardi, M. Mey and C. Giuntini,
   Chest tomography by gamma camera and external gamma source,
   J. Nucl. Med. 19:94, 1978.
3. C. Giuntini, R. Guzzardi, M. Pistolesi, M. Mey, and S. Solfanelli,
   Evaluation of a system for 90° Compton scattering in lung tomography, Progr. Resp. Res. 11:76, 1979.

COMPUTER ANALYSIS OF DYNAMIC KRYPTON 81m LUNG SCANS IN INFANCY

Charles Y. C. Wong and Michael Silverman

Hammersmith Hospital
London W12 OHS
England

Krypton 81m has been used recently for ventilation and perfusion in paediatrics. The low radiation dosage, short scanning time, and appropriate energy for high resolution make it particularly suitable for young uncooperative patients. We have developed a simple technique for studying regional lung function using $^{81}Kr^m$ and evaluated it in relation to other techniques for assessing lung function.

For a ventilation scan, infants are placed on top of the gamma camera and inhale air enriched with krypton gas while a mini-computer records one second histograms of radioactivity counts throughout the procedure. At steady state an image is recorded photographically by accumulation of oscilloscope display. After about 30 seconds, the supply of krypton is abruptly stopped for the acquisition of wash-out data. The perfusion scan is done with the same set-up except krypton is dissolved in 5% dextrose and infused via a peripheral vein.

Due to the high ventilatory turnover (ventilation per unit lung volume) in infancy, the ventilation scan steady state image represents distribution of ventilated volume. Thus the image can only be interpreted when regional ventilatory turnover is known from the wash-out analysis. For the perfusion scan, the steady state counts depend on the arrival of krypton by perfusion and departure of krypton by ventilatory turnover of perfused alveoli. These two factors can be resolved by calculating regional perfusion, corrected for differences in ventilatory turnover.

Regional ventilatory turnover rates are calculated by estimation of the slope of the initial part of the wash-out log-activity/time curve. In view of the fast physical decay of krypton, the initial slope reflects both the fast and slow compartments. A basic pro-

45

gramme has been written to do this heavy analysis, for pre-determined
regions of interest. After decay correction of the wash-out, the dynamic scan data is displayed to check visually that it is fit for
analysis. The steady state counts are summed over the 10 seconds
prior to the onset of the wash-out. An exponential regression routine then calculates the initial wash-out slope. The "initial" range
of the slope is defined by stripping the curve from the onset of the
wash-out to the point where the coefficient of variation of the regression line of the slope is minimal while including a substantial
number of data points. Simple calculations then provide numerical
values for regional ventilation, ventilated volume, perfusion and
normalised V/Q ratios, as well as providing an estimate of confidence
limits of these parameters.

A combined follow-up study with plethysmography, chest X-ray,
and clinical follow-up of twenty newborn infants was undertaken to
evaluate the method and clinical value of krypton scans. Results
clearly show that the steady state image alone is of limited value.
Examination of the wash-out curves, and their mathematical analysis,
yielded insight into lung pathophysiology.

In combination with other observations, krypton dynamic lung
scans provided a more complete picture of the progress and severity
of post-neonatal lung disease in this group.

REFERENCES

1. F. Fazio and T. Jones, 1975, Assessment of regional ventilation
    by continuous inhalation of radioactive krypton-81m. British
    Medical Journal, 3:673-676.
2. G. Giofetta, T. A. Pratt, J. M. B. Hughes, 1978, Regional pulmon-
    ary perfusion assessed with continuous intravenous infusion of
    $81_{Kr}m$: a comparison with 99Tc^m-macroaggregates (99Tc^m-HAM).
    Journal of Nuclear Medicine, 19:1126-1130.
3. G. Ciofetta, M. Silverman and M. Hughes, Quantitative approach to
    the study of regional lung function in children using krypton-
    81m. Brit. Med. Radiol. (In Press).

COMPARISON OF METHODS TO CALCULATE REGIONAL LUNG VOLUME AND

SPECIFIC ALVEOLAR VENTILATION FROM XENON-133 WASH-IN CURVES

Th. W. van der Mark, H. Beekhuis, R. Peset and
M. G. Woldring
Lung Function Laboratory and Nuclear Medicine Dept.
University Hospital
Groningen
The Netherlands

Wash-in curves for physiologically inert gases are usually described by a mono-exponential function of the type $N(t) = N_\infty(1-e^{-kt})$. N_∞ is related to lung volume and K is the specific alveolar ventilation, or alveolar ventilation per unit lung volume. Determination of the specific alveolar ventilation therefore recess takes an estimate of K from the measured curve. Compared with other gases Xenon-133 is fairly soluble in blood (blood-gas partition coefficient $\lambda = 0.15$[2]). This gives rise to multiexponential curves[3].

Multi-exponential analysis of wash-in curves is difficult to perform. Since we are mainly interested in the lung parameters N_∞ and K, we treat the exponentials describing the uptake of Xenon-133 in blood and tissue, as pertubations, by expanding them in a power series.

For the determination of N and K a function of the form

$$N(t) = N_\infty (1-e^{-kt}) + a_0 + a_1t + a_2t^2 \qquad (1)$$

has been fitted to the measured curve by a non-linear least squares technique. The extra quadratic term in this function describes the background.

The results of this method have been compared with the results of several conventional methods of estimating K: the determination of the half-time, the height-over-area method, and determination of the mean transit time, both in simulated curves and in curves from patients. This comparison shows that within the "normal" range of K for spontaneous breathing ($K \sim 1\text{-}2 \ min^{-1}$) the differences between the various methods was within 15%. Large deviation however occurs for low and high values of k when conventional methods are used. For these values of k the proposed method clearly gives the best result, in particular in reduced alveolar ventilation such as occurs in chronic airflow obstruction.

However, the more elaborate least-squares fitting method requires much computer time, which makes it inappropriate for large data-sets and for applications on small computers. From our results it appears that the half-time method performed best of all conventional methods. The value of N_∞ found with the half-time method is often too low, because of incomplete wash-in. This can be adjusted by taking into account the estimated half-time. The result is a new value for N_∞ and a new value for $t-\frac{1}{2}$. This procedure can be repeated and it can be proven mathematically that this procedure converges, provided that the value of $t-\frac{1}{2}$ is smaller than the wash-in time.

Also we can estimate the value of the background after the wash-out, and extrapolate this value to the wash-in part of the curve, using average values found for the coefficients in the quadratic function (eq. 1), or using the data of Süsskind and others.

These techniques result in a good estimate of k and N_∞, and can be run on small computers. Errors are less than 6% in the range $0.3 \text{ min}^{-1} < k < 2.5 \text{ min}^{-1}$. This method is therefore suitable when measuring regional lung function in patients with chronic airflow obstruction.

REFERENCES

1. S. S. Kety, The theory and applications of the exchange of inert gases at the lungs and tissues, Pharmacol. Rev. 3:1-41 (1951).
2. A. M. Anderson and J. Ladefoged, Relationship between hematocrit. and solubility of Xenon-133 in blood. J. Pharm. Sci. 54:1684-1685 (1965).
3. H. Süsskind, H. L. Atkins, H. Cohn et al., Whole body retention of Radio-Xenon, J. Nucl. Med. 18:462-471 (1977).

IDENTIFICATION METHODS FOR ANALYSIS OF XENON-133 WASH-OUT CURVES:

A CRITICAL COMPARATIVE STUDY

V. Brusasco, A. Tiano, R. Astengo and S. Valenti

Clinica Tisiologica e Malattie dell'Apparato
Respiratorio, Università di Genova
Genoa, Italy

Xenon-133 is the radioactive tracer gas most widely used for studying regional ventilation; however, a serious limitation is constituted by its relatively high solubility in blood and tissues.

The purpose of this paper is to investigate the adequacy of two methods for the computer analysis of regional xenon-133 wash-out curves, paying attention to the effects of tissues saturation.

METHODS

Xenon-133 wash-out curves were recorded for 3 minutes at the mouth and on the chest of five sitting healthy subjects breathing at constant tidal volumes and frequencies, after 3 minutes of equilibration on a closed circuit system. This procedure was then repeated without time interval between the end of wash-out and the beginning of the next rebreathing, so that tissues saturation was allowed to increase.

The first identification method assumes a discrete-time stochastic model of the autoregressive type:

$$y(t) = \sum_{i=1}^{n} a_1 y(t-i) + \varepsilon(t) \qquad\qquad t = 0,1,2,\ldots \qquad (1)$$

where $y(t)$ represents count rate at time t and $\varepsilon(t)$ is a sequence of stationary white noise signals. The aim of identification is to determine from a set of observed data $(y(t) = t = 0,1,2,\ldots, n)$ the order n and the parameters $(a_1, a_2, \ldots, a_n)$ more consistent with data. For this purpose a Least Squares Parameter Estimation of models of increasing order (n = 1,2,...,m) is carried out, while the optimum order determination is recursively accomplished by minimising the Final Prediction Error criterion.

The second identification method assumes a statistical continuous compartment model:

$$y(t) = \int_0^\infty G(k)e^{-kt}dk \qquad\qquad (2)$$

where G(k) represents the distribution function of lung constant rates, k, and is computed by calculation of the distribution moments through the Edgeworth-Cramer series expansion.[1]

The mean value of chest wall background count obtained after pneumectomy in five patients was introduced to correct wash-out curves: this value was 0.3 min^{-1} ± 0.033 S.E. which is consistent with results by Jones[2] and Matthews and Dollery.[3] We estimated the weight of this component to be about 30% of total count rates at the beginning of wash-out.

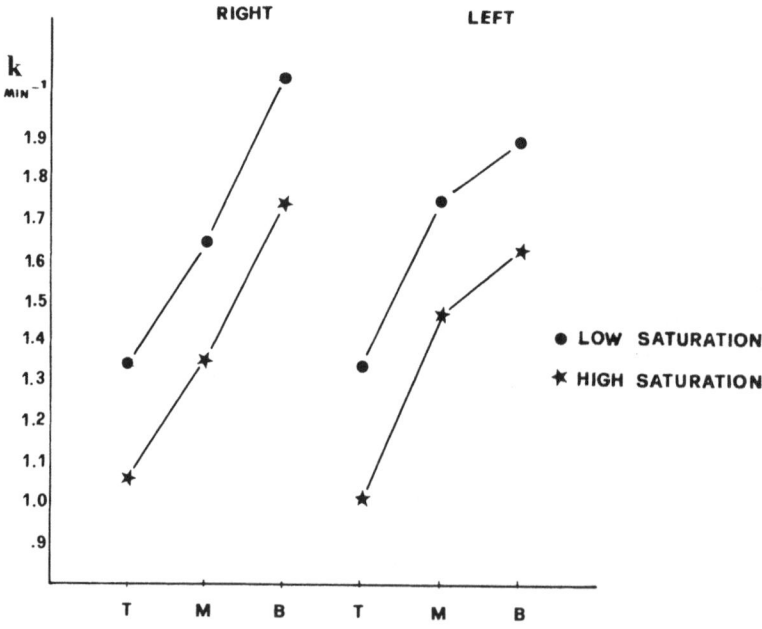

Fig. 1. Mean values of constant rates (discrete-time model with .3 weight assigned to chest wall) from two subjects at very different levels of saturation: low corresponds to trials with average residual count rate about 10% after 3 minutes of wash-out; high to trials with average residual count rate of about 30%. Distance top (T) to bottom (B) = 15 cm.

Table 1. Cost Functions (FPE x 10^{-2}; Mean Values ± S.E.)

weight	ALL REGIONS Order 1	2	3	continuous	MOUTH CURVES Order 1	2	3	continuous
0	.1368	.1330	.1368	.1373	.1882	.1906	.1987	.1802
	±.0119	±.0128	±.0133	±.0117	±.0352	±.0373	±.0392	±.0254
.3	.1363	.1305	.1341		.2004	.1972	.2046	
	±.0106	±.0119	±.0125		±.0336	±.0397	±.0410	

RESULTS

In most regional curves minimum cost function was achieved by a discrete-time model in its second order, nevertheless in many cases the first order was more adequate to fit experimental data. The introduction of the estimated chest wall compartment often improved the fitness of regional curves but worsened that of the mouth curves (Table 1).

Constant rates of wash-outs at mouth converted into Darling's alveolar dilution factors are consistent with spirographically measured parameters, but this is not the case for values from the continuous-time model (Table 2). The K values inferred from first order equations or from the slow compartment of the second order show an apico-basal gradient consistent with predicted difference of lung expansion, while very fast compartments (12.22   8.3 SE. min^{-1}) resulted occasionally and do not show vertical orientation. Regional constant rates were slower than those at the mouth but this difference was almost completely eliminated by taking into account the chest wall wash-out rate.

Progressive saturation reduced all constant rates but did not affect the apico-basal gradient.

CONCLUSIONS

Our discrete-time model seems adequate for computer analysis of regional Xenon-133 wash-out curves, being scarcely affected by

Table 2. Mean Values of Constant Rates (min^{-1}) ± S.E.

Weight Run	Mouth	right lung			left lung		
		top	middle	bottom	top	middle	bottom
A -discrete time							
1	1.706	1.035	1.382	1.603	1.044	1.362	1.630
	±.221	±.070	±.115	±.155	±.082	±.111	±.187
2	1.668	1.026	1.307	1.525	.946	1.319	1.484
	±.310	±.058	±.131	±.198	±.069	±.126	±.200
1		1.296	1.539	1.810	1.298	1.634	1.832
		±.089	±.135	±.159	±.115	±.126	±.200
2		1.228	1.444	1.734	1.167	1.508	1.618
		±.072	±.142	±.241	±.085	±.130	±.209
B -continuous time							
1	3.170	1.520	2.742	3.448	1.550	2.224	3.020
	±.562	±.376	±.567	±.758	±.497	±.389	±.604
2	2.923	1.571	2.009	3.126	1.488	2.280	3.194
	±.610	±.254	±.341	±.876	±.347	±.601	±.870

Alveolar dilution factors (from spirographic and antropometric data) = .834 ± 0.20. Frequency = 10.5 ± 1.0 min^{-1}. Distance top to bottom = 15 cm.

tissues wash-out while the continuous-time model seems to be subject to the experimental error in the first few breaths of wash-out, as suggested also by the occasional occurrence of very fast compartments in discrete-time results.

These analyses are feasible through a low cost microcomputer (TRS-80) programmed in Basic Level II and Assembler languages, interfaced with a multidetector system.

REFERENCES

1. G. Lamedica, V. Brusasco, A. Tiano and R. Ramoino, Analysis of
   nitrogen multi-breath wash-out curves through a statistical
   approach. In press on Respiration.
2. H. B. Jones, Respiratory system: nitrogen elimination. In:
   "Medical Physics vol. II" O. Glasser ed., The Year Book
   Publishers, Chicago (1950).
3. C. M. E. Matthews and C. T. Dollery. Interpretation of xenon-133
   lung wash-in and wash-out curves using an analogue computer.
   Clin. Sci. 28:573 (1965).

CHAPTER 3

THE LUNG FUNCTION LABORATORY SPIROMETRY
MECHANICS AND RESPIRATION

THE FLOW-VOLUME-MEASUREMENT IN COMPUTER-SUPPORTED LUNG FUNCTION ANALYSIS

Christoph Riemasch-Becker and Klaus Stosseck

Institut für Anaesthesiologie der Johannes Gutenberg
Universität Mainz
Mainz
Bundesrepublik Deuschland

For some years computers have been applied more and more in pulmonary function laboratories. On the one hand the medico-technical industry could connect older diagnostic systems with new calculators, on the other hand completely computerised lung function desks have been developed.

In the hospital pulmonary function laboratory of our Institute of Anaesthesiology in Mainz we have been working since March 1979 with the complete Pulmonary System Desk No. 47120 A by Hewlett-Packard, which permits the measurement and the calculation of the values of ventilation, distribution and diffusion. The corresponding diagrams are printed additionally. Three pneumotachographs in combination with gas-analysers for nitrogen, carbonmonoxide and helium enable us to perform tests of Forced Vital Capacity (FVC), spirometry, maximal voluntary ventilation (MVV), nitrogen wash-out by single or multiple breath method (NWs, NWM) and diffusion capacity by single breath method (DCOsb). After comparison with reference values, which can be chosen between two sets of predicted value equations[1,2,3] the criteria in Table 1 are used to produce the impression statements. Pre-bronchodilator results are compared to predicted values and post-bronchodilator results are compared to the pre-bronchodilator values.

Within these tests of lung function the Forced Vital Capacity with the flow-volume curve is one of the most important measurements to verify restrictive or obstructive lung disease. There are up to 19 different parameters (Table 2) of volume, flow and time values obtained from this test.

Table 1. Impression Statements

	Measured value	Impression statement
PRE	% predicted 80%	normal
BRONCHO-	% predicted 65%	slightly reduced
DILA-	% predicted 50%	moderately reduced
TORS	% predicted 50%	severely reduced
POST	% of best pre 125%	markedly improved
BRONCHO-	% of best pre 115%	improved
DILA-	% of best pre 105%	not clearly improved
TORS	All other	not improved

Table 2. Forced Vital Capacity Report Values

Volume (L)		Flow (L/sec)	Time (sec)
FVC	FEV2	FEF .2-1.2	FVC TIME
FEV.5	FEV2/FVC	FEF 25-75	MET
FEV.5/FVC	FEV3	FEF 75-85	
FEV1	FEV3/FVC	PF	
FEV1/FVC	IVC	MEF 50%	
		MEF 75%	
		MIF 50%	

In 500 pulmonary function test reports the Maximal Expiratory Flow at 50% Forced Vital Capacity - MEF 50% - was found occasionally, and the Maximal Expiratory Flow at 75% Forced Vital Capacity - MEF 75% - was found frequently reduced without any clinical relevance. So it may be asked, whether these two flowparameters are more of theoretical than of clinical interest.

With respect to these reductions we tested two groups of 30 male and 30 female healthy physicians, nurses and medical technicians. In the male group the MEF 50% was reduced in 17, the MEF 75% in 25

Table 3. MEF 50% and 75% Reductions in 60 Healthy Adults

Reduction of	male (n = 30)	female (n = 30)
MEF 50%	17 (56.6%)	13 (43.3%)
MEF 75%	25 (83.3%)	16 (53.3%)

persons, in the female group there were reductions in 13 and 16 persons respectively (Table 3).

Thus in the male group the incidence of reduced flow was higher than in the female group: while the maximal expiratory flow at 50% Forced Vital Capacity was reduced by about 50% in both groups, the incidence of reduction in Maximal Expiratory Flow at 75% FVC was 30% higher in the male than in the female group. Furthermore, we found that the degree of reduction in Maximal Expiratory Flow at 75% FVC depends significantly on the duration time which is needed to exhale Forced Vital Capacity (Fig. 1): the relation between the reduction of the MEF 75% and the FVC time is shown in this diagram. The

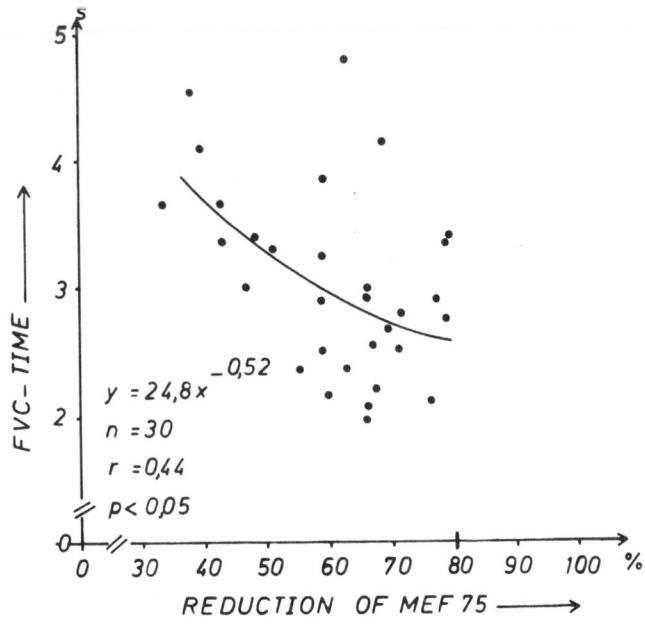

Fig. 1. The relation between the reduction of the MEF 75% and the FVC time.

reduction is given in percentages (less than 80%; see the impression statements) the FVC time in seconds. A significant correlation was found by using a potential curve fitting. It can be seen that the degree of reduction increases with FVC. If the FVC exceeds 3 seconds, the MEF 75% is reduced for more than 50%. Thus, wrong pathological results may occur, for the FVC depends on individual cooperation too.

Three examples of good or doubtful cooperation are given. In the first one of a test person with exact cooperation the FVC is done in less than 3 seconds and there is no reduction in the parameters. In the second one the colleague's cooperation was not satisfactory. While the flow-volume pattern is nearly regular, many flow parameters, MEF 50% and 75% included, are reduced. FVC is higher than expected and the FVC is prolonged to about 7 seconds. In contrast to these two examples the third one with small airway disease is able to exhale the FVC in normal time (in about 3 seconds), and the flow-parameters are reduced in the same quantity as in example two.

In spite of the calculated "guide lines" before patient FVC-test manoeuvre (Table 4), it seems suitable to use additional criteria, for instance the FVC time, or its product with the Forced Vital Capacity in the form of a trivial "Control Factor" (COF). By calculating this "Control Factor" for the two groups of tested persons, we found the following values (Table 5).

Table 4. Predicted Values Printed Before
Patient FVC-test Manoeuvre

FVC - FEV 1 - %FEV1/FVC - FEF 25 - 75 PF

Table 5. "Control Factor" for Normal Adults

FVC * Time	Male (n = 30)		Female (n = 30)	
COF	25	5	18	3

Firstly this factor can help to identify real pathologic results. Secondly this allows the detection of inadvertently retarded test manoeuvres and a reduction in the amount of unsuccessful results.

In any case we should include all lung function parameters of the FVC-test for a correct diagnosis. In patients with lung disease it is recommended that the MEF 50% and 75% reduction is interpreted and correlated with other volume, flow and time parameters in order to verify restrictive or obstructive lung disease. In this way too, it is possible to distinguish whether obstruction is located in bigger or smaller airways.

It is possible to adapt the lung function test programme within certain limits at the physician's personal request by changing the software configuration, for instance some test-parameters can be excluded. This might lead to a more comfortable application, especially if controversial values are omitted. In our opinion, however, the advantage of a computerised test programme is to have the use of a complete set of possible lung function parameters in combination with the plotted diagram for their interpretation.

REFERENCES

1. R. J. Knudsen, R. C. Slatin, M. D. Lebowitz and B. Burrows, The Maximal Expiratory Flow-Volume Curve Normal Standards, Variability, and Effects, Am. Rev. of Resp. Dis., 112:587 (1976).
2. J. F. Morris, A. Koski and L. C. Johnson, Spirometric Standards for Healthy Nonsmoking Adults, Am. Rev. of Resp. Dis. 103:57 (1971).
3. J. F. Morris, A. Koski and J. D. Breese, Normal Values and Evaluation of Forced End-Expiratory Flow, Am. Rev. of Resp. Dis. 111:755 (1975).

COMPARISON OF ALGORITHMS FOR THE DIAGNOSIS OF OBSTRUCTIVE AND

RESTRICTIVE LUNG FUNCTION IMPAIRMENT

W. Arossa, S. Spinaci, A. Carosso, G. Forconi and
E. Concina

Consorzio Provinciale Antitubercolare, CNR Unit
Sub Project BCPO, Lgo Dora Savona 26, 10152 Turin, Italy

INTRODUCTION

Published validations of computer reporting program have only been limited to a review of generated reports,[1] without any definition of the type of error.

We therefore set about studying the efficiency, and the relative error; of six diagnostic algorithms to distinguish between obstruction and restriction when lung function tests are below a defined normal limit.

METHOD

A random sample of 537 patients was studied. Each patient had at least a chest X-ray, a standard CECA questionnaire, spirometry (VC, FEV1, FEV1%. TLC) and a clinical examination. We performed broncho-dilator and codiffusion tests in patients below the normal limit in one of the above tests. Predictive values were derived from the equations of CECA[2] for lung volumes, and of Cotes[3] for diffusion; the normal limits have been set at the reference value minus two standard deviations. All functional diagnoses were comprehensively defined by two physicians, as follows:

normal:	all functional tests within normal limits,
obstruction:	impairment sequential X-ray, questionnaire and clinical examination suggestive of emphysema, asthma or chronic bronchitis,
restriction:	X-ray clinical evidence of fibrosis, radiological tubercolosis, obesity,
mixed:	any combination of the previous impairments.

The physicians' diagnosis was considered as the reference for
every comparison with computer generated diagnoses, performed using
six different algorithms (Table 1).

Table 1. Definition of Algorithms

Algorithm	Obstruction	Restriction
A 1	FEV1/Pred <81%	VC/Pred <83%
A 2	FEV1/Pred <81%	TLC/Pred <78%
A 3	FEV1/VC Pred <87%	VC/Pred <83%
A 4	FEV1/VC Pred <87%	TLC/Pred <78%
A 5	The same as A 4	

If $\frac{FEV1/VC}{Pred}$ and TLC/Pred are normal and VC/Pred <75% (i.e. 3 standard
   deviations),
If TLC single breath He/TLC Plethysomgrafic - >0,9, diffusion
   normal, no change after bronchodilators = restriction,
If TLCsbHe/TLCplet <0,9, significative change after bronchodilators
   = obstruction,
Any other condition = insufficient data, check clinical data

A 6   The same as A 5
If $\frac{FEV1/VC}{Pred}$ <87% and VC/Pred <65% = mixed impairment

RESULTS

The sensitivity and spedificity of four diagnostic markers (defined as the ability) to detect the physicians' diagnosis of lung
function impairment were:

sensitivity: VC 97%; TLC 54%; FEV1 96%; FEV 1% 98%
specificity: VC 19%; TLC 99%; FEV1 91%; FEV 1% 99%

The results of data processing are reported in Table 2.

Table 2.   Results of Data Processing

Diagnosis	Correct	A1	A2	A3	A4	A5	A6
Obstruction	144	87*	161	82	150	150	125
Restriction	18	27*	4	43	14	15	15
Mixed	15	88*	14	72*	4*	4*	29*
Normal	360	335	358	340	369	360	360
Total	537	537	537	537	537	529	529

* p <0,05

The program A5 accounts for 98.5% of all the patients because
in eight cases it requested clinical data. This program made errors
in seven patients with airways obstruction, TLC within normal limits
but reduced VC and an associated restriction. The best cut-off level
of VC, used to eliminate these errors, proved to be 65% of the reference value with an overestimation of mixed impairments (Table 2;
column A6).

DISCUSSION

The use of the computer to report lung function tests depends on
the assumption that: "if clinical diagnosis is to be made on a
logical basis, the results of each test should be stated objectively
and in isolation, and only then applied to the other information already available" as stated by Geddes et al.[1]

The main problem is: which algorithm achieves the best agreement
with the physicians' diagnosis when lung function tests cannot be interpreted without any other clinical data.

Our study shows that the best markers are FEV1% and TLC ratio.
VC and FEV1, on the other hand, are more sensitive to any kind of
lung function impairment; thus may then lead to gross misinterpreta-
tions. The percentage of correct diagnoses, using the first two
markers, is 97%. The algorithm fails to reach the 100% because of
errors mainly in patients with a mixed impairment.

We have found that in these conditions the only way to reduce
errors is to reject an automated diagnosis, and to generate a

request for further clinical data, because all the algorithms checked do not change significantly the definition rate.

Our results confirm that it is possible to identify a unique index of lung function impairment and the need for clinical data can be limited to a small number of patients.

REFERENCES

1. D. M. Geddes, M. Green and P. A. Emerson, Comparison of reports on lung function tests made by chest physicians with those made by a simple computer program, Thorax, 33:257 (1980).
2. W. Bolt, D. Brille, M. Cara, G. Coppée, A. Houberechts, F. Lavenne, P. Sadoul, E. Sartorelli and O. Zorn, "Aide-mémoire pour la pratique de l'examen de la fonction ventilatoire par la spirographie," CECA, Luxembourg, (1971).
3. J. E. Cotes, "Lung function," Blackwell Scientific Publications, London, (1979).

AN AUTOMATED SYSTEM FOR THE MEASUREMENT OF AIRWAYS RESISTANCE

LUNG VOLUME AND FLOW-VOLUME LOOPS

P. J. Chowienczyk, P. J. Rees, T. J. H. Clark

Respiratory Function Unit
2nd Floor New Guy's House, Guy's Hospital
London Bridge, London SE1 9RI
England

OBJECTIVES

We have developed a computerised plethysmograph system for the measurement of airways resistance, lung volumes and flow volume loops. Conventional techniques for determining airways resistance and lung volumes give results which have large intra-subject variability. In developing this system, therefore, our aims have been to develop algorithms which give objective and reproducible measurements of airways resistance and lung volumes.

METHOD

The system hardware consists of a constant volume plethysmograph, a digital computer and a high resolution graphics terminal. Plethysmograph pressure, mouth pressure and flow are fed to the computer via a 10-bit analogue to digital converter.

The manoeuvre performed by the patient may be divided into three stages. Stage one consists of a two second period of panting at about 120 breaths per minute; plethysmograph pressure and flow are sampled for the calculation of airways resistance. Stage two consists of a further period of panting for 1.5 seconds against a closed mouth shutter; thoracic gas volume is calculated from mouth pressure and plethysmograph pressure. In stage three the mouth-shutter opens and the patient inspires to total lung capacity and then forcibly expires to residual lung volume. This allows the lung volume and flow-volume loop to be obtained from the flow signal (which is integrated to give volume).

67

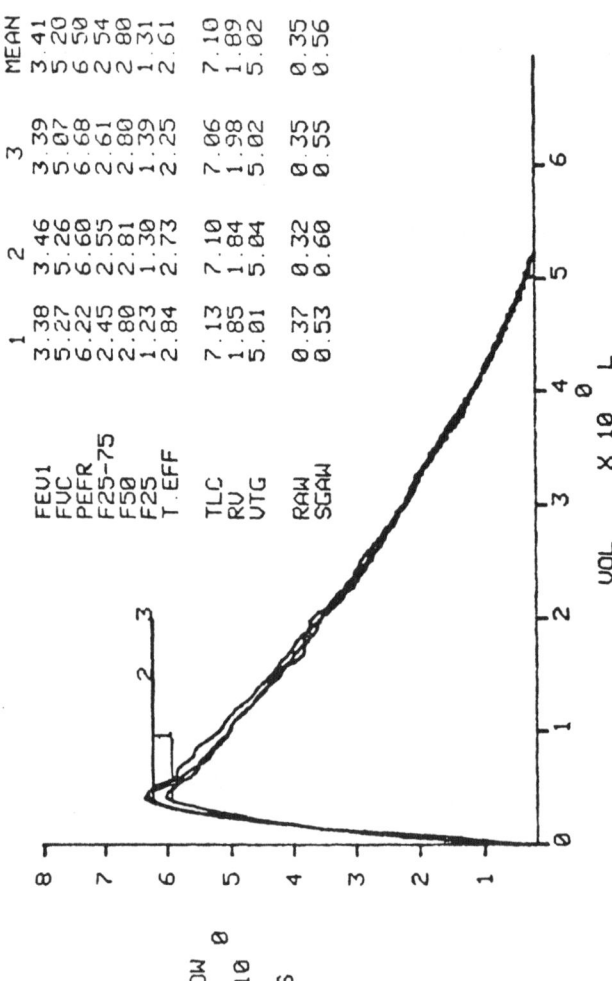

Fig. 1. The hard copy output by the system after three manoeuvres have been performed by the patient. The flow-volume loops are shown together with measurements derived during the first, second and third stages of the manoeuvres: FEV1; forced expiratory volume in one second, FVC; forced vital capacity, PEFR; peak expiratory flow rate, F25-75; mean flow from 25% of the vital capacity to 75%, F50; flow at 50% of the vital capacity, F25; flow at 25% of the vital capacity, T.EFF; effective time constant (or mean transit time), TLC; total lung capacity, RV; residual lung volume, VTG: thoracic gas volume, RAW; airways resistance, SGAW; specific airways conductance.

Airways resistance is determined from the pressure and flow waveforms, sampled during the first stage of the manoeuvre, by using a least squares technique to fit, to these waveforms, sine waves of the same frequency. Resistance can then be calculated from the amplitudes and phase relation of the fitted sine waves by using the formula:

Raw = K. (PM/VM). cosQ

where PM is the amplitude of the pressure sine wave, $\dot{V}M$ is the amplitude of the flow sine wave, Q is the phase angle between the two and K a constant.

Thoracic gas volume is calculated from mouth pressure and plethysmograph pressure signals sampled during the second stage of the manoeuvre. The sampled data points are devided into groups for which the rate of change of mouth pressure is within set limits. A regression line is fitted to each group and the average of the slopes of these lines is used to calculate thoracic gas volume. This procedure prevents bias from artifacts, such as glottic closure, which occur in subjects with poor technique.

RESULTS

Measurements of airways resistance and thoracic gas volume derived by the system correlate well with those obtained by conventional methods. The coefficient of variation of specific airways conductance is less than 10% and that of total lung capacity in the order of 5%. Results are obtained within 30 seconds of starting the manoeuvre and the procedure can be repeated immediately. Figure 1 shows the hard copy obtainable from the system after three manoeuvres have been performed.

CONCLUSIONS

Our system provides measurements of airways resistance, lung volumes and flow-volume loops from a single manoeuvre. Results can be obtained within 30 seconds of starting the manoeuvre. The use of a computer has made possible more objective and reproducible measurements than can be obtained by hand.

ANALYSIS OF BODY PLETHYSMOGRAPHIC PRESSURE-FLOW-LOOPS BY DIGITAL COMPUTER

Dieter Heise

Medizinische Universitätsklinik und Poliklinik
Abt. Pneumonologie
D-6650 Homburg/Saar
West-Germany

In obstructed patients body plethysmography often yields non-linear pressure-flow-curves. The loop-formations are due to trapped air (hysteresis), central obstruction (s-shape), variable bronchial geometry (8-shape) and high transmural pressures during expiration (club-shape). Our method allows for the numerical analysis of all loops.

While the box door is closed, the patient breathes normally and then with slightly increased frequency. These manoeuvres are interrupted by shutter closure. Immediately the box door is opened, the patient performs slow vital capacity twice and forced vital capacity up to three times. Flow, volume, box pressure and mouth pressure are transferred to a disk file at the end of each manoeuvre. Storing of artifacts is avoided by displaying appropriate diagrams on a x-y storage oscilloscope.

Systematic errors of body plethysmography are reduced by pre-processing the stored data. Alveolar pressure is calculated from box pressure by use of a formula[1] which allows for changing alveolar volume and compression of alveolar gas due to obstruction. The influences of temperature and humidity on box pressure, flow and volume are corrected. The lung volume is derived as absolute volume by starting the integration of flow at a thoracic gas volume calculated from the slope of the shutter curve.[2]

The calculated alveolar pressure can be split into a resistive and a compressive part, if there is hysteresis because of trapped air. Since the compressive part p_{alv}^c can be assumed to be proportional to the volume changes during the breathing cycle, it can be subtracted from the calculated alveolar pressure p_{alv}:

71

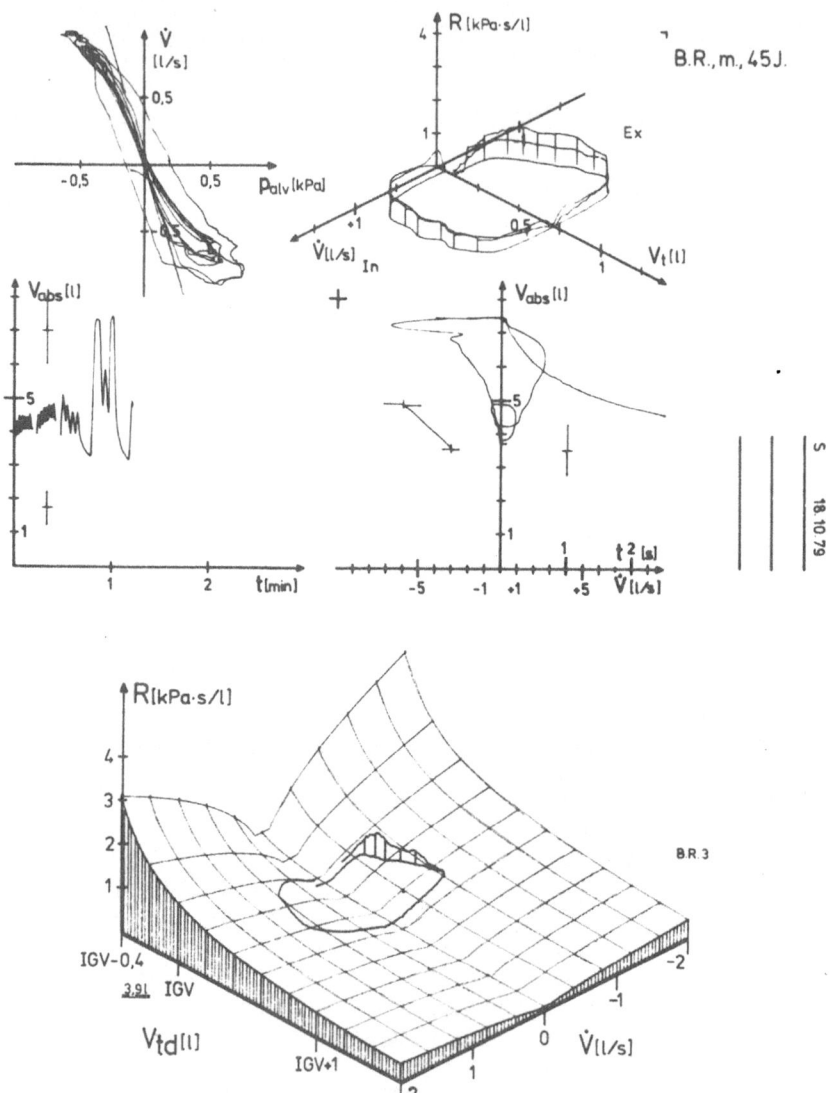

Fig. 1.   Graphical representation of measurement obtained by com-
          bination of body plethysmography and spirometry.

$$p^r_{alv} = p_{alv} - k_{ta}\Delta V$$

The resistive part p^r_{alv} is used to calculate the airway resistance. The model equation

$$R = R_0 \cdot (\dot{V}/\dot{V}_0)^a \cdot ((V_{abs} - RV)/ERV_0)^B$$

describes the simultaneous dependence of airway resistance on flow and volume by the flow exponent a (0<a<1) and the volume exponent B (-2<B<0). It holds for inspiration only because expiration may be handicapped by high transmural pressures. To estimate the expiratory handicap, we first calculate R_0, a and B by two-dimensional logarithmic regression on inspiratory values of flow, volume and resistance. Then we compute a theoretical course of resistance by use of these parameters and expiratory flows and volumes. The differences between the measured and the computed resistance are the result of the expiratory handicap.[3] Figure 1 shows the graphical representation of the results. Where applicable, the plots contain the predicted values and their statistical deviations. The static lung volumes can be derived from absolute lung volume vs. time. The forced vital capacity manoeuvre is plotted twice: volume vs. time and volume vs. flow. The pressure-flow loop shows original alveolar pressure and its resistive part if available. Finally there are two three-dimensional plots which demonstrate the dependency of airway resistance on flow and volume. The first contains vertical time marks (0.1 s), the second extrapolates the flow-volume dependency by use of a grid. The region of expiratory handicap protrudes vertically from the surface of the grid.

The method yields useful information on the features of breathing mechanics especially in chronic obstructive lung disease.

REFERENCES

1. D. Heise, Das Prinzip der Molzahlen in der Theorie der Ganzkörper-Plethysmographie - Atemwegs - und lungenkrankheiten, 5:133-135 (1979).
2. D. Heise, Computer-aided measurements in body plethysmography. Progress in Respiration Research 11:179-187 (1979).
3. D. Heise, Neue Interpretationsmöglichkeiten bei der Ganzkörper Plethysmographie, Forbildung in Thoraxkrankheiten, 9:231-238 (1980).

NON-INVASIVE ALVEOLAR PRESSURE/FLOW PATTERN DETERMINATION BY

COMPUTERIZED PLETHYSMOGRAPHY

M. E. Perry, R. W. Zimmerer, R. J. Browning

Pulmonary Function Laboratory/Clinical Investigation
Service, Fitzsimons Army Medical Center
Aurora, Colorado 80045, U.S.A.

In the past measurement of alveolar pressure patterns has
required the insertion of uncomfortable esophageal balloon catheters.
We have developed a plethysmographic method to measure alveolar
pressure during forced expirations. We use a DEC PDP 11/10 computer
to sample plethysmograph pressure and pneumotach flow at .002 second
intervals. After correcting for heat exchange effects, data are displayed as flow/volume and flow/alveolar pressure plots.

Although usually ignored, heat transfer in and out of the sealed
gas system has significant effects on plethysmograph pressure and can
introduce considerable error in any subsequent calculation of
alveolar pressure. If these effects are disregarded, errors in
excess of 20 centimeters of water are commonplace over the terminal
portion of expiration when alveolar pressure is returning towards
zero.

Our method assumes isothermal conditions within the alveoli and
conducting airways. Applying Boyles law:

$$\frac{P_h + \Delta P(t) - 47}{P_b - 47} = \frac{V(t)}{V(t) - V_c(t)} \tag{1}$$

Where P_b is barometric pressure and $\Delta P(t)$ is alveolar pressure
(gauge). $V(t)$ is the uncompressed lung gas volume deduced from integrated pneumatic flow after correction for non-linearity. Resting
end-tidal volume is the reference point for flow integration and is
assumed equal to the previous known FRC. $V_c(t)$ is the volume by
which the lung is compressed and is equal to the volume change
sensed by the plethysmograph pressure transducer after corrections
for plethysmograph "drift" and pneumotach heat exchange have been

75

M. E. PERRY ET AL.

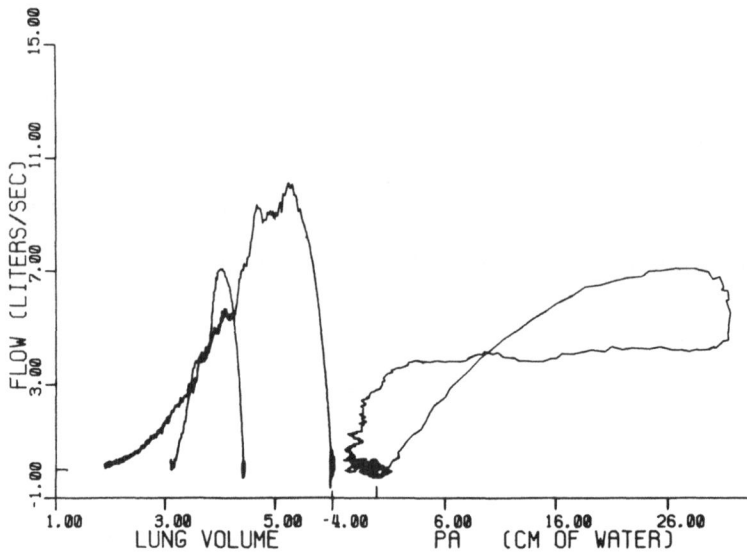

Fig. 1.   Computer plot of a "puff" effort shown on the left super-
          imposed on the maximum flow/volume loop. The correspond-
          ing alveolar pressure is shown on the right.

made. The standard calibration procedure allows the changes in
plethysmograph pressure to be expressed as changes in plethysmograph
volume. We have found experimentally a 2% loss of exhaled volume
presumably due to cooling and desaturation as exhaled gas passes
through the heated pneumotach. At the end of the forced manoeuvre,
the overall box volume change due to pneumatic effect is calculated
from the total exhaled volume. Any remaining difference in plethysmograph volume between that immediately before and that immediately
after the manoeuvre is attributed to drift, which is the unpredict-
able result of heat transfer through the gas system from the
patient's body and the surrounding environment. This is then distributed equally over the sample intervals since the drift can be
assumed relatively constant during the short duration of the
manoeuvre, typically less than 0.5 seconds. After the drift correction, plethysmograph volumes at each sampling interval are further
corrected for pneumotach effect proportional to the volume exhaled
over each interval. After these corrections, the box volume change
at each interval is equal to $V_c(t)$ and can be substituted into
equation 1. After $\Delta P(t)$ is computed the calculated flow-dependent
pneumotach back pressure is subtracted resulting in instantaneous
alveolar pressure referenced to mouth pressure.

     During gentle puff manoeuvres the relationship between flow
and pressure is linear. With puffs of greater intensity a curvilinear relation emerges consistent with a more turbulent flow
regime. Puffs of sufficient intensity to pierce the maximum flow/

volume envelope causes the pressure/flow relation to appear as an
open loop with a prominent pressure-independent constant flow portion. When a series of puffs or coughs are performed only those
individual manoeuvres that pierce the flow volume envelope develop
this pattern.

Our data demonstrate relationships previously known to occur
only through the use of more invasive techniques. The non-invasive
nature of our method allows for greater subject acceptance of
alveolar pressure measurement. We are presently searching for
correlation between the degree of obstructive airway impairment and
the critical pressure required to obtain flow maximums. This may
lead to a clinical useful diagnostic procedure.

A NEW METHOD FOR ESTIMATING CHANGES IN BRONCHOMOTOR TONE USING THE APPLE MICROCOMPUTER

J. R. Lehane, C. Jordan, J. P. Royston and J. G. Jones

Division of Anaesthesia, Clinical Research Centre
Watford Road, Harrow, Middlesex
England

Studies of the effects of therapy on bronchomotor activity in man are complicated by the fact that airways resistance (Raw) may be markedly influenced by changes in lung volume (V_L). Specific airways conductance (SGaw), derived from either the ratio of conductance (1/Raw) to thoracic gas volume, or the slope of a plot of conductance against volume, may be used as an index of bronchomotor tone. We have adapted the forced airflow oscillation method for measuring resistance to determine SGaw in patients when the body plethysmographic method is unsuitable (e.g. during anaesthesia or IPPV) or unavailable.

An active or passive deflation of 3 litre/min. from end-inspiration to near residual volume was performed. During this manoeuvre a sinusoidal airflow (3 Hz, peak flow 0.55 l/sec) was imposed on the airway. Respiratory resistance (Rrs) was calculated continuously from measurements of airway pressure and airflow and plotted against lung volume. Analysis of the resistance/volume plots obtained in 20 anaesthetised intubated patients, and in 11 awake volunteers confirmed the previously described hyperbolic relationship between Rrs and V_L. It also revealed that extrapolation of these curves to infinite lung volume yielded an asymptotic value for Rrs that was not zero but was a positive value whose magnitude varied. Plots of conductance against V_L were consequently curvilinear and hyperbolic and it was concluded that SGaw as determined by the plethysmographic methods was not independent of lung volume. In order to produce a linear conductance/volume plot it was necessary to subtract this asymptotic resistance from the resistance data before plotting its reciprocal against lung volume. This was performed using a microcomputer aided hyperbolic curve-fitting analysis which determined the asymptotic values and derived

79

derived Sgaw. The Apple II microcomputer sampled approximately 50 resistance-volume coordinates automatically during each lung deflation, performed the curve-fitting analysis and then presented the results as a graphical display of the data points and the fitted curve, together with a table of the parameters. The method consistently provided a very good fit to the data and produced linear conductance/volume plots in both anaesthetised patients and awake subjects.

To assess the effectiveness of the method in detecting changes in bronchomotor tone in conscious volunteers we studied the effects of 1) bronchodilator aerosol, rimiterol (Riker Laboratories Ltd), 2) a histamine aerosol followed by rimiterol aerosol or placebo aerosol. Healthy, non-smoking volunteers were studied, seven in trial 1 and four in trial 2.

In trial 1, there was a significant increase in SGaw to 138% of control ($P<0.05$) following one dose of rimiterol (0.2 mg), indicating bronchodilatation. There was a further increase in SGaw to 151% of control ($P<0.01$) following a further dose of rimiterol (0.4 mg). In trial 2, SGaw was significantly depressed to 30% of control after four breaths of a 2% histamine aerosol, indicating bronchoconstriction ($P<0.05$). SGaw returned to control values within 20 min of 0.4 mg rimiterol but remained depressed after placebo aerosol. The difference between rimiterol and placebo was statistically significant ($P<0.05$, paired t test).

The method was therefore sensitive to changes in bronchomotor tone and may prove to be a useful technique in the assessment of the effects of therapy in conscious or mechanically ventilated patients.

REFERENCES

1. W. A. Briscoe and A. B. Dubois, The relationship between airway resistance, airway conductance and lung volume in subjects of different age and body size, J. Clin. Invest. 37:1279 (1958).
2. H. Watt, R. C. Zimmerman, I. R. Peters and W. J. Sullivan, Direct write out of total respiratory resistance, J. Appl. Physiology. 28:675 (1970).

COMPUTER ANALYSIS OF BREATHING PATTERNS

R. G. Loudon, A. G. Leitch, T. H. Ridgway, J. K. Walsh
and M. Kramer

Department of Internal Medicine, Chemistry, Psychiatry
and the Sleep Disorder Centre, University of Cincinnati
Cincinnati, Ohio 45267 USA

Breath-by-breath patterns of ventilation have been analysed
using a micro-computer, and a program package developed specifically
for this purpose. Previous studies by others have categorised the
duration of inspiration (T_I) and of expiration (T_E), and the "duty
cycle," (the proportion of the respiratory period which is occupied
by inspiration (T_I/T_{TOT}) ratio). Cyclical variation has been noted.
Findings have varied, depending on the method used to record ventilation and on the circumstances at the time of measurement (awake or
asleep, rest or exercise, metabolic load). Breathing patterns during
sleep have been the subject of considerable attention recently.[1]
Observations in animals suggest that changes in breathing with each
sleep stage result from changing central programs.[2] The clinical
importance of breathing patterns during sleep has been demonstrated
or suspended in the sudden infant death syndrome, in the sleep apnea
syndromes, in chronic mountain sickness, in unexpected nocturnal
asthma deaths, and in patients with chronic ventilatory failure.[3]
Our objective was to develop programs which would allow the analysis
of breathing patterns over long periods, for example eight hours, in
the sleep laboratory eight to ten thousand breaths.

We recorded ventilation signals using nasal and oral thermistors,
a chest pneumograph, or a two-coil string-vest inductance plethysmograph (Respitrace, Ambulatory Monitoring, Ardsley, New York). The
signals were recorded on FM analog magnetic tape and analysed off-
line. The first step in analysis was to measure the duration of
successive inspirations and expirations. A program loop with a 20
millisecond cycle type received the signal, tested for zero crossing
(thermistor signal) or for maximum or minimum (plethysmograph) and
incremented a counter until the beginning of inspiration or

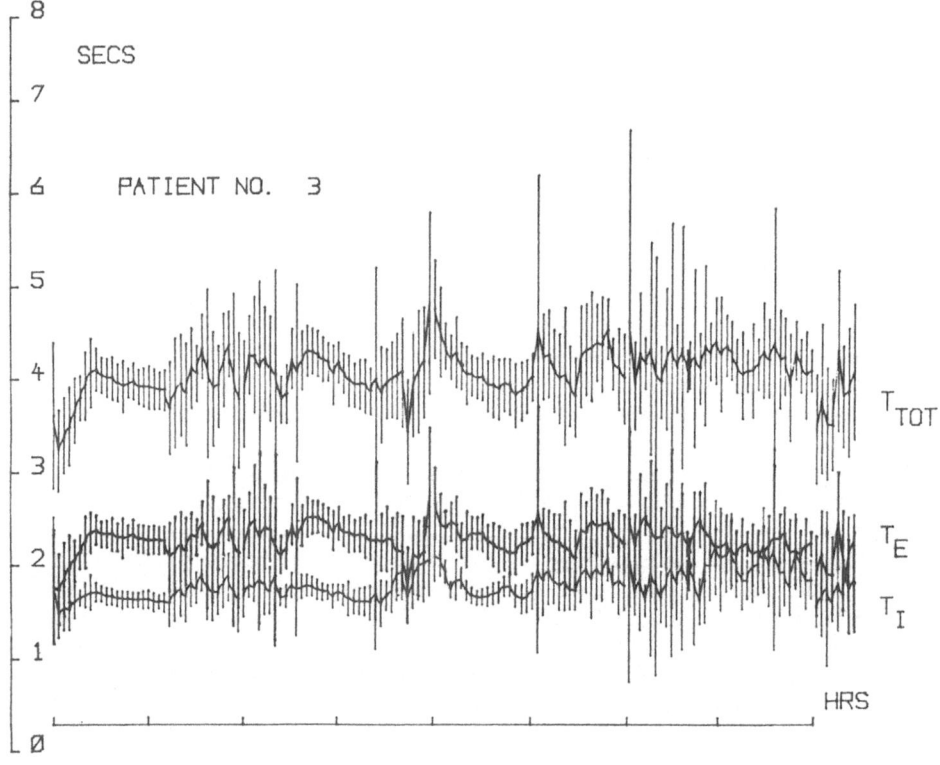

Fig. 1.

expiration was detected. The count was then stored and the counter reset. After 600 breaths the analog recorder was stopped, the 1200 T_I and T_E values were transferred to cassette, and the analog tape recorder under computer control was backed up one turn and another block of 600 breaths was analysed. This analysis takes eight hours, but requires no supervision.

Once the blocks of data are on cassette, they are printed out and edited if necessary. Overlaps from block to block can be deleted and "glitches" caused by transients removed by an automatic editing program, and the repeat printout checked against the polygraph record. Any further editing necessary is facilitated by a manual editing program which presents the data on the screen and allows various manipulations, correction, and recording of the edited data. Blocks of data can be split into smaller blocks, and sections chosen to correspond to sleep stages, with the help of other programs.

The next program used calculates means and standard deviations for blocks of data, for T_I, T_E, and T_{TOT}, and records the results on cassette, printing them out at the same time; another program then calculates the autocorrelation function to seek periodicity in blocks of data, again printing out the results and recording them on cassette. The final group of programs plots the sequence of T_I. T_E, and T_{TOT} means and standard deviation for an eight-hour period (Fig. 1), plots T_I against T_E breath-by-breath for blocks of chosen size, and plots autocorrelation functions sequentially or as required.

Our results support Orem's suggestion that differences in breathing during sleep derive from changing central programs. An interesting new finding is that sequences of T_I show periodicity not seen in the corresponding T_E sequence. We plan to apply this method of analysis to patients with a variety of respiratory abnormalities.

REFERENCES

1. E. A. Philipson, Control of breathing during sleep. Amer Rev. Respir. Dis. 118:909 (1978).
2. J. Orem, A. Netick and W. C. Dement, Breathing during sleep and wakefulness in the cat. Respir. Physiol. 30:265 (1977).
3. A. G. Leitch, L. J. Clancy, R. J. E. Leggett, P. Tweeddale, P. Dawson, and J. I. Evans, Arterial blood gas tensions, hydrogen ion and electroencephalogram during sleep in patients with chronic ventilatory failure. Thorax 31:730 (1976).

AUTOMATIC DATA PROCESSING IN THE CLINICAL LUNGFUNCTION LABORATORY

A. F. M. Verbraak, J. M. Bogaard, F. R. C. Jansen,
and A. Versprille

Pathophysiological laboratory
Department of Pulmonary Diseases
University Hospital "Dijkzigt"
Rotterdam
The Netherlands

Approximately 15,000 lung function measurements are performed yearly on about 4,000 patients. The off-line automatic data processing (ADP) of the hand measured values is 4 years in use now and has resulted in:
- a capacity increase of 20%,
- a greater accuracy in calculation which implies an improvement of the quality of comparisons between patients' data and reference values, and
- a standardisation of the lung-function report.

On-line ADP is partly realised at the moment. The main advantage is the immediate availability of the data during a patient's programme besides a further increase of capacity. Moreover, on-line ADP results in a reduction of measurements for the detection of maximal performed values and in an efficient investigation programme of lung function tests based on the results of the immediately foregoing measurements and the expected pathology.

The main requirements for the ADP are:
- input of patient's data, done at the administration desk as well as in the different investigation rooms
- graphic representation for checking the results of the measurements
- data storage, important for the generation of the patient report, recalculations and research analysis
- a maximum flexibility during the measurements in order to make the programme fit the patient

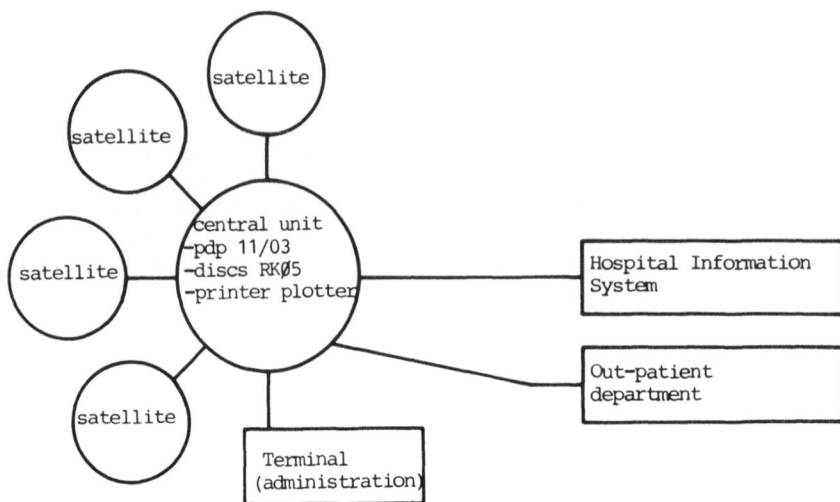

Fig. 1.   The multi-processor system in its final configuration. The
          connections between the central unit and respectively the
          out-patient clinic, the Hospital Information System and the
          satellites in the investigation rooms are indicated. The
          expensive peripherals, like the mass storage device and the
          electrostatic printer/plotter, are only necessary at one
          place: the central unit.

        - results of measurements in the different rooms to be enlisted
          to a final report
        - suitability for programme development and clinical research
        - integration with the Hospital Information System (HIS)
        - support of the screening measurements in the outpatient clinic
        - a more extended signal analysis
        - production of one final lung-function report of all tests in
          one patient, even when performed in different investigation
          rooms
        - simple handling by the lung function technician

Because no commercial system could fulfil our requirements, a local
development appeared to be necessary.

    One of the main problems of the on-line ADP is to structure the
programmes in such a way that a maximum flexibility is obtained with
respect to the interaction patient-computer technician. For the communication between computer and technician we use function keys. The
interaction of the technician with the computer is reduced by use of
pattern recognition techniques. A log-file of the successive activi-
ties of the technician during the measurements enables, together with
the stored signals, an off-line recalculation. In the programme the
possibility to correct earlier decisions is implied. Selections from

the stored values can be made from series of measurements, in between
as well as at the end. Finally, the on-line and off-line collected
data are integrated to a final lung function report. At the moment
the on-line processing of the flow-volume curves and the measurements
of static and dynamic lung volumes by spirometer are tested in routine use at one working place.

     For the realisation of the on-line ADP we have chosen the multi-
processor approach (Fig. 1). The main benefits of this set-up are:
     - the relatively simple computers
     - the parallel processing of the measurements in different rooms
     - the interchangeability of the system components, and
     - the extension of the system according to the patient-load.
The general function of the central unit, a PDP-11/03 with two disc
drives and an electrostatic printer plotter are:
     - communication between the HIS, the outpatients' department and
       the satellites
     - the storage of programmes and patient data and the printing of
       the reports.
The satellites consist of a micro-processor (PDP-11/23), a switch
panel, a graphic display and the lung function equipment. This
system handles all on-line measurements. Moreover each satellite can
be used for programme development. For one of the satellites we have
planned a disc drive in order to function as a background system and
in order to use it for special tasks.

CHAPTER 4

THE LUNG FUNCTION LABORATORY (GAS EXCHANGE)

ON-LINE PROCESSING OF EXPIRATORY pCO_2-CURVES FOR DIAGNOSIS OF PULMONARY EMPHYSEMA

U. Smidt and H. Worth

Bethanien Hospital
Moers
West Germany

Pulmonary emphysema is defined as an enlargement of the air spaces distal to the terminal bronchioles. From anatomic studies it is known that emphysema does not begin in the alveoli, but in the airways of generation 16-21. During an expiration the gas coming from those generations can be expected after the first 100-200 ml, i.e. after the dead space gas.

This next part of expired gas is the part of mixed air, labelled as "phase II" of an expiratory partial pressure curve and followed by phase III, the "alveolar gas." From clinical studies we know that phase II in emphysema is much longer and comprises more volume than in healthy subjects. This difference is easily detected in time plots of expiratory partial pressure curves, but it is still more pronounced in plots of partial pressure vs. expired volume (Fig. 1). When the same tidal volume is exhaled more rapidly, the time plots will be compressed, but the volume plots remain unchanged. On the other hand, when changing tidal volumes are exhaled in the same time, the shape of the volume plot will change considerably, whereas the time plots do not show much change. However, the volume of the part of mixed air cannot be predetermined by the finally expired volume, because the part of mixed air is already exhaled before the total duration and volume of expiration are fixed. The deciding variable for the volume of mixed air is the previously inspired volume (V_{IN}). Therefore we relate the volume of mixed air (V_M) to V_{IN}.

The main problem from the theoretical and practical point of view was the definition of V_M.

After various approaches we decided to take the volume expired between 25% and 50% of the inspiratory end-tidal concentration difference of the respective gas.

91

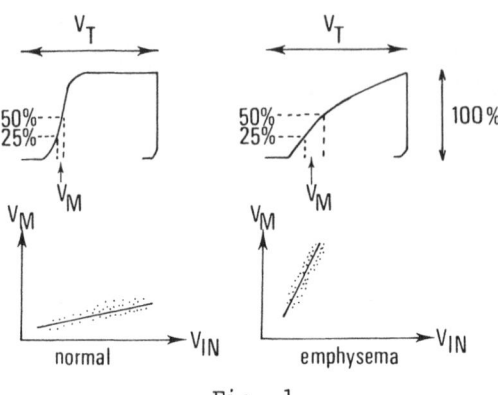

Fig. 1.

This definition has the advantage that it is independent of absolute pressures. Practical experience has shown that this definition also eliminates the influence of hyperventilation and physical
exercise. Moreover no gas calibration is necessary. For the volume
no calibration is necessary because both V_{IN} and V_M are volumes and
their relationship is of interest. The relation V_M to V_{IN} is
strongly linear, in healthy subjects about 25-40 ml/l, in patients
with advanced emphysema up to 100 ml/l.

Regarding the micro-processor we sample the flow signal and the
pCO_2 signal with 50 Hz and use a resolution of about 256 for 1 l/s
and 4 for 1 mm Hg pCO_2. Recognition of the start of inspiration and
expiration is based on the crossing of a zero band with a width of
± 1/32 l/sec. The condition for acceptance of a breath for evaluation is that its inspiratory and expiratory volume are not more than
4-times smaller or larger than the previous breath. Inspiratory pCO_2
is not defined as the smallest, but as the most frequent pCO_2 during
inspiration. End-tidal pCO_2 is defined as the highest pCO_2 during an
expiration.

The 25% and the 50% points are defined as the first sample ex-
ceeding these percentages. When, for instance, these points are 32%
and 54%, so that the difference is 22% instead of 25%, the volume
between these points is multiplied by 25/22. This correction reduces the scatter considerably. After completion of these calculations and plotting the respective point in the V_M - V_{IN} diagram, all
data of the breath are discarded, except V_{IN} and V_{EX}, which will

still be used for comparison with the size of the next breath. The
micro-processor has 1K (12 bit) for flow samples and 1K for pCO_2
samples, so that breathing frequencies down to 3/min are theoretically possible. The programme for sampling, calculations and plotting is shorter than 1K, so that a total of 3 K is sufficient.

AUTOMATIC COMPUTATION OF SINGLE BREATH NITROGEN TEST

P. Pisani, P. Paoletti, C. Marchesi, E. Fornai,
P. Fazzi, F. Di Pede, G. Pistelli, C. Giuntini

CNR Clinical Physiology Institute and 2nd Medical
Clinic, University of Pisa
Pisa, Italy*

Single breath nitrogen test (SBN) provides a considerable amount of information about different indices of lung function:[1] closing volume (CV), slope of alveolar plateau ($\Delta N_2\%lt$), total lung capacity (TLC) and derived indexes (functional residual capacity and residual volume). CV and $\Delta N_2\%lt$ are considered suitable to detect early stages of airways obstruction. Thus they have been proposed for epidemiological studies.

Automatic determination of CV remains a problem because of difficulty in detecting the take off of phase 4 on the N_2 curve ("closing point"). The contour of the N_2 curve is variable from patient to patient, and within the same patient is strongly dependent on the way the manoeuvre is performed. The different shapes of the N_2 curve are empirically grouped in five classes: 1. curves with phase 3 and closing point clearly detectable; 2. curves with superimposed noise from cardiogenic oscillations at the onset of phase 4; 3. curves with different levels on the alveolar plateau; 4. curves with continuous rather than abrupt slope change in the region of the closing point, and 5. curves with steep alveolar plateau, with or without detectable closing point. Current computer programs[2] detect a closing point only when very evident.

---

*This study was supported by National Research Council, Rome, Italy. Project on Preventive Medicine, Subproject on Chronic Obstructive Lung Disease.

We propose a different approach[3] based on the selection of a
limited number of possible closing points. The computer analysis of
the N_2 curve includes two successive steps. A "significant edge" of
the curve occurs when the derivative calculated on the points belong-
ing to it stays with the same sign at least for a 300 ms interval;
the longer the interval, the more significant the edge. In the first
step a searching interval (SI) containing the most significant edges,
to be further analysed, is defined. In the second step a regression
line is calculated over the main portion of the alveolar plateau.
Finally each onset of the edges, up to a maximum of 5, within the SI
having the largest areas delimited by the extrapolated regression
line, is considered for selection as possible closing point. The N_2
curve is displayed on a CRT with markers on the possible closing
points, a special marker indicates the most significant edge selected
by the computer (Fig. 1).

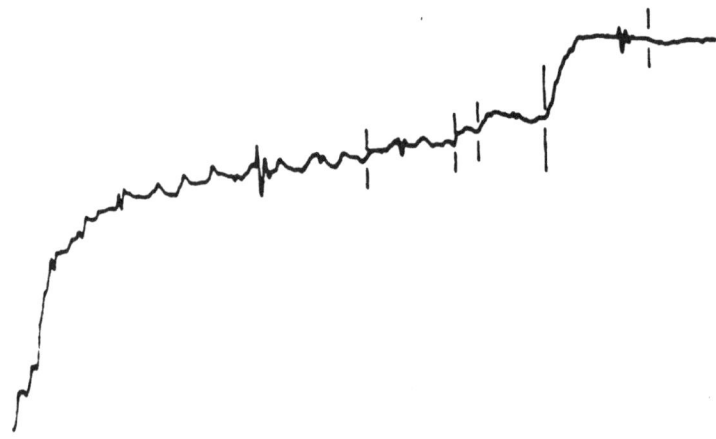

Fig. 1.

At this stage the operator may interact with the computer. If
he does, he may choose only one out of the possible points. Thus
standardisation may be achieved even in this case despite operator
variability. He will have to choose a point in the region of the
closing point, because of the constraint of choosing only one closing
point.

54 manual and computer determinations of CV, $\Delta N_2\%/1t$ and TLC
were compared in 20 subjects selected to yield all the defined 5
classes of N_2 curve. Closing point was selected entirely by computer
in 42 instances, in 7 closing points was chosen by the operator among
the possible closing points and in 5 curves the operator judged inadequate the computer selection and chose his own closing point.

Manual and computer determinations (49 CV, 54 $\Delta N_2\%/It$, 54 TLC) were
compared by three different statistical methods: the mean difference,
the regression analysis and the randomised block variance analysis.
Results show good agreement between manual and computer determinations.

Further improvements are planned to achieve a higher degree of
standardisation to encompass the potential of the large epidemiological application of this method. The ΔN_2 determination depends,
in the current methods, on the reliability of the closing point
detection. The independent ΔN_2 determination together with a fully
automated method for closing point detection allows a higher accuracy
of the measurements in the epidemiological surveys.

REFERENCES

1. A. S. Buist, The single breath nitrogen test, New Engl. J. Med.
    28:438 (1975).
2. N. Craven, G. Sidwall, P. West, D. S. McCarthy, R. M. Cherniack,
    Computer Analysis of the Single Breath Nitrogen Wash-out
    curve, Am. Rev. Resp. Dis. 113:445 (1976).
3. P. Paoletti, E. Fornai, P. Pisani, C. Marchesi, P. Fazzi, G.
    Pistelli, C. Giuntini, Automatic procedures for pulmonary
    function tests in epidemiology, in: "Proceedings of Inter-
    national Conference on Recent Advances in Biomedical Engineer-
    ing Society," Published in Great Britain by the Biological
    Engineering Society.

AUTOMATED MEASUREMENT OF SINGLE-BREATH DIFFUSING-CAPACITY (TLSB) IN ROUTINE LUNG-FUNCTION TESTING

M. Heitz, M. Küng, A. Perruchoud, C. Kopp, H. Herzog

Department of Internal Medicine, Division of Respiratory
Disease, Cantonal Hospital
CH-4031 Basel, Switzerland

The measurement of transfer factor by the single breath test with carbon monoxide (TL_{SB}) has become well accepted. Introduction of automated equipment for valve-control[1] and a calculator or computer for computations enable the test to be undertaken in routine pulmonary function laboratory.

We evaluated the extent to which transfer factor could replace our current tests for assessment of diffusion disturbance - a steady state diffusing capacity (DLss) and trend of arterial oxygen tension (PaO2) and alveolo-arterial oxygen gradient (P_AO_2 -PaO$_2$) during exercise.

MATERIAL AND METHODS

In 70 patients with sarcoidosis, interstital fibrosis, or allergic alveolitis we compared transfer factor measured by the single-breath technique at rest with DLss, PaO2 and alveolo-arterial oxygen gradient ($AaDO_2$) under steady state bicycle exercise (40 W) in the supine position. The test is performed according to the technique of Ogilvie[2] with a commercially available valve and sampling system. The valves are controlled by 4 programmable read-only memories (ROM) which use signals of airflow, volume and a quartz-clock as input. These signals as well as those of helium and CO-analysers are also fed into a PDP 11-34 computer, but the latter is not used to control the valves of the sampling system. Effective time of breath holding is calculated according to Cotes.[3] No correction was made for carboxyhemoglobin. The time required for one complete measurement is 4-5 minutes.

The mean coefficient of variation of 5 measurements in each of 5 normal subjects was 0.038 (range 0.03 - 0.05).

RESULTS AND DISCUSSION

a) All patients (n = 70). A significant correlation existed between TL_{SB} and DL_{SS} (r = 0.68). This was also found for TL_{SB} and PaO_2 at the two exercise-levels 40 watts (r = 0.53) and 100 watts (r = 0.674).

b) Patients with diffusion-impairment (n = 25). Impaired diffusion was arbitrarily defined by a decrease in PaO_2 of 3 mm mercury or more from rest to exercise and an increase in ($PAo2-PaO2$). In 25 patients correlation between TL_{SB} and DL_{SS} was good (r = 0.82). Correlation between TL_{SB}/m^2 and PaO_2 (40 watt) was only r = 0.45, p = 0.05 and was less than the one between DL_{SS}/m^2 and PaO_2 (r = 0.64). No correlation existed between $TL_{SB}/m2$ and P_AO_2 - PaO2) (40 watt) but was significant for $DL_{SS}/m2$ and $AaDO_2$ (r = 0.70).

c) Patients with treatment and follow-up examination. In 5 of 6 patients treated because of impaired diffusion, PaO_2 (40 watt) and TL_{SB} changed with therapy in the same, in 1 patient in the opposite direction.

Our results show that TL_{SB} is a useful screening measurement. A normal TL_{SB} and an abnormal PaO_2 under exercise are unlikely. TL_{SB} is probably also suitable to assess effects of therapy. In patients with an impaired diffusion no conclusions can be made from TL_{SB} about the level of hypoxemia (PaO_2) during exercise.

REFERENCES

1. E. A. Gaensler and A. A. Smith, Attachment for automated single breath diffusing capacity measurement, Chest, 63:136-145.(1973).
2. C. M. Ogilvie, R. E. Forster, W. S. Blakemore and J. W. Morton, A standardised breath holding technique for the clinical measurement of the diffusing capacity of the lung for carbon monoxide, J. Clin. Invest. 36:1-17 (1957).
3. J. E. Cotes, "Lung Function," Blackwell Scientific Publications, London (1965).

BLOOD GAS ANALYSIS - GRAPHICAL DATA REPRESENTATION BY COMPUTER

APPLICATION

E. Voigt

Institute of Anaesthesiology of the University of
Tübingen, GFR

A rapid analysis of acid-base parameters and blood gases is necessary and desirable for many purposes. In many blood gas analyses computers are integrated, delivering the data more precisely than it is possible by slide rules or nomograms. Beyond the quick availability of the several parameters due to therapeutic management of severe disturbances of the acid base equilibrium the graphical representation of the data is desirable for instance in education or for recognition of more complex connections.

A system is described, consisting of a blood gas analysis (Gas-Check-AVL 933), a programmable desktop calculator (HP 9825), two data lines to the cardiac surgery unit and to the intensive care unit, and a four colour X-Y-plotter (HP 9872).

For running the routine programme the measured values for pO2, pCO2 and pH, further the hematocrit and the patient's temperature are fed in "off line" to the calculator. The measured values, the corrected values for actual temperature and the derived parameters are immediately printed out. After verifying the data, the transmission of the measured and derived data via the two data lines is performed. The transmission and outlining of all data require only 6 seconds.

In addition to this numerical data print out, a graphical representation of several parameters is possible on a four colour X-Y-plotter. The routine plot compromises the pH/HCO3 nomogram including the marked actual situation, the direction of partial metabolic or respiratory compensation of the disturbances towards pH 7.4, and a computer diagnosis of the actual disturbance according to Suero.[2]

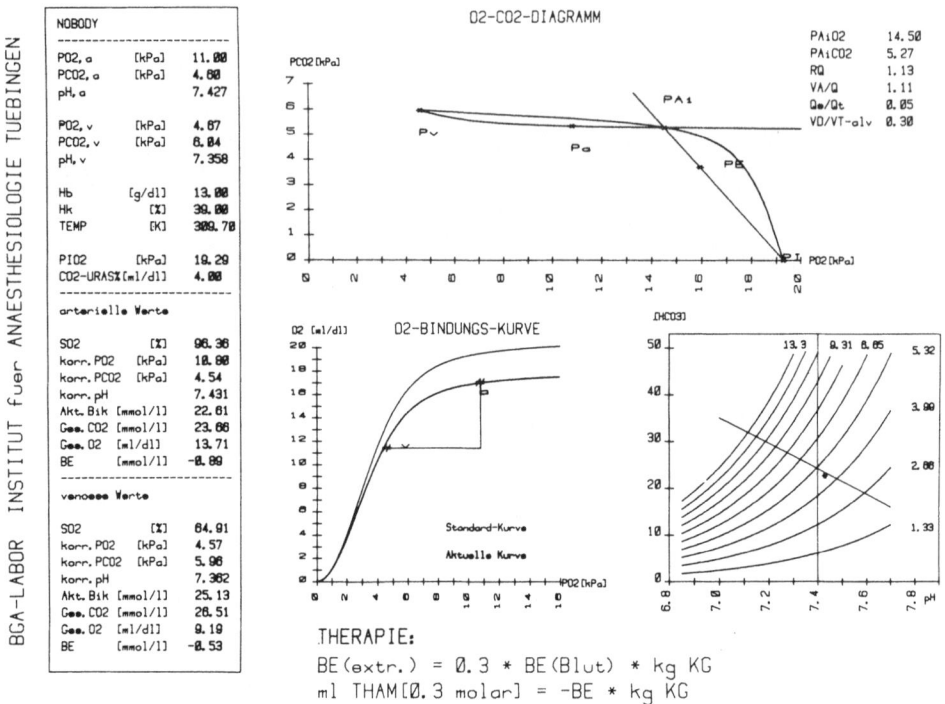

Fig. 1. The whole acid-base and blood gas status, consisting of O_2-
CO_2-diagram, O_2-dissociation curve and acid-base-nomogram
with all measured and derived parameters.

If a central venous blood probe is available in combination with
an arterial probe, the actual and standard oxygen dissociation curves
can be plotted with marked pO_2, 50 and the arterio-venous oxygen difference for recognition of peripheral ischaemic, anaemic or hypoxic
anoxia.

With the initial parameters of an arterio-centralvenous blood
probe, the inspiratory O_2- and end-expiratory CO_2-concentration the
O_2-CO_2-diagram can be plotted. From this diagram further values are
calculated such as ventilation/perfusion-ratio ($\dot{V}A/Q$), venous admixture (Qs/Qt) and alveolar dead space ventilation (VD/VT). In
combination with this O_2-CO_2-diagram the standard and actual oxygen
dissociation curves with arterio-venous oxygen difference and the
acid-base nomogram are plotted (Fig. 1). The several programmes are
written in HPL (Hewlett/Packard Language) and in the present form of
the normal routine the programme includes up to 5108 bytes. The
several programmes are stored on a high speed data cartridge
(capacity up to 2500 bytes). At each demand they can be read in the
calculator's memory, consisting of 6844 bytes, with a speed of 14300

bytes/s. The time needed for plotting demands on the programme
varies between 1 to 8 min. The algorithms used in the different
programmes base on the formulas according to Thomas.[1]

REFERENCES

1. L. J. Thomas, jr., Algorithms for selected blood acid base and
      blood gas calculations, J. Appl. Physiol. 33:154 (1972).
2. J. T. Suero, Computer interpretation of acid-base data, Clin.
      Biochem. 3:151 (1970).

A COMPUTER PROGRAM FOR THE LUNG-FUNCTION LABORATORY

Per Malmberg and Kjell Rasmundson

The Department of Clinical Physiology
University Hospital
S-750 14 Uppsala, Sweden*

INTRODUCTION

A complete lung-function evaluation comprises many tests, such
as measurement of lung volumes, airway resistance, ventilatory flow,
gas transfer, gas distribution, closing volume and lung mechanics.
Many tests are performed several times before and after broncho-
dilation and of the total number of measured data many exceed 150.

The complete lung-function program offers increased validity in
conclusions regarding pulmonary function and is especially valuable
in the detection of early occupational lung disease. The demand for
the complete test program is increasing, however, computing results
and preparing the report is very time consuming even with the aid of
table top computers (more than 1 hour/patient). Therefore a computer
program was developed on a mini computer.

OBJECTIVES OF THE COMPUTER PROGRAM

The program was designed to cover all steps in the complete lung-
function analysis, from on-line or off-line input of data to output
of the final report. The program was designed to allow easy operation, allowing on-line editing of input data and to accept input
from a great variety of recording instruments. Finally it was constructed to allow easy modification and expansion of the program.

---

*Supported by grants from the Swedish Work Environment Fund, Project
75171.

RESULTS AND DISCUSSION

The program uses a PDP1135 mini computer with RSX11-M operating system (time share). The mini computer has an A/D converter (AR11), a 2.5 mbyte disc (RKO5), a graphic terminal (VT55), a printer (Decwriter) with Swedish alphabet and a printer plotter (Versatec). In the future a PDP1123 mini computer with 5 mbyte discs (RLO1) and a VT100 graphic terminal will be used. The program uses overlay structure reducing core memory requirement to 26 words.

The program has a modular structure, each (input) module corresponding to a lung-function test "module." The modules have similar structure. After input of data, the input signal is checked by displaying the input signal with measuring points on the graphic terminal together with results in absolute values and in relation to predicted values and results of previous measurements. Data from the present measurement may now be rejected or stored on disc, and previous measurement data may be deleted from the disc. Then more data can be entered from the same patient, from new or previous patients or the program operation can be stopped and resumed at a later time. The communication language is Swedish.

In the evaluation program, unsuitable data are automatically rejected and appropriate means or maximum values are computed. It is possible to override bad data at any time by inputting correct data. The final protocols are printed in condensed tables and interesting curves are plotted as formatted graphs.

At present several input modules for analogue on-line input of data exist (different spirometric tests) and other are in preparation (closing volume). These modules also have an off-line manual input counterpart as back up and in addition there are several modules for manual off-line input of sample tests. Preliminary tests with the program indicate that it reduces the time of each investigation, increases precision and enhances readability and understanding of the lung-function protocol.

CHAPTER 5

NEW TECHNIQUES IN THE LABORATORY AND
INTENSIVE CARE UNIT

THE SOLITARY PULMONARY NODULE - DECISION ANALYSIS

R. Kunstätter, N. Wolkove, H. Kreisman, C. Cohen
and H. Frank

Pulmonary Division, Department of Medicine
Sir Mortimer B. Davis Jewish General Hospital and
McGill University
Montreal, Canada

A solitary pulmonary nodule is a finding on chest X-ray showing a single circumscribed mass in the lung. This finding is usually not associated with any other abnormality or any symptoms and is often first noticed on routine examination. Many disease processes, particularly lung cancer, tuberculosis, and fungal infections of the lung may cause such a nodule. Though biopsy with a fine needle may be used to establish the nature of the nodule, often this technique is unavailable, or yields no definite result. In this case thoracotomy is needed to obtain a diagnosis. The advantages of surgery are that it assures a definite diagnosis and is also the best treatment for lung cancer which presents as a solitary nodule. However, the potential long-term benefit of surgery must be balanced against both the immediate risk of the operation and the advantage of not operating on those patients who have a benign nodule. At present the physician must decide the case before surgery. Without the benefit of precise quantitative data or the formal techniques to apply such data to an individual patient.

The object of our research is to provide a more formal and explicit framework within which this decision may be considered. We have designed a decision analysis model and prepared a computer programme which facilitates both the application of this model to individual patients, and the modification of this model to allow for future refinements.

The model may be conveniently represented as a decision tree (Fig. 1). The first node of the tree is a decision node (drawn as a square) and represents the choice between surgery and no surgery.

Clearly, such decision nodes represent events over which we have full
control and which require an active choice on our part. The second
node of the tree is a chance node (drawn as a circle) and represents
the possibility of peri-operative death for those patients who undergo
surgery. Chance nodes, unlike decision nodes, represent events of
nature over which we have no control but whose outcomes we can predict
with probabilities. Here, M is the probability of peri-operative
death and 1-M is therefore the probability that the patient survives
surgery. Stated in a more familiar way, M is the mortality rate for
this type of surgery. The next level of the tree reflects the fact
that both those patients who survive surgery and those on whom no
surgery has been performed, may or may not have had cancer. C is the
probability that the solitary pulmonary nodule is cancer and 1-C is
the probability that it is not. The last node indicates that patients
who undergo surgery and have cancer may or may not have a resectable
lesion. R is the probability that a lesion is resectable and 1-R is
the probability that it is not. Thus, R is the resectability rate of
malignant solitary pulmonary nodules.

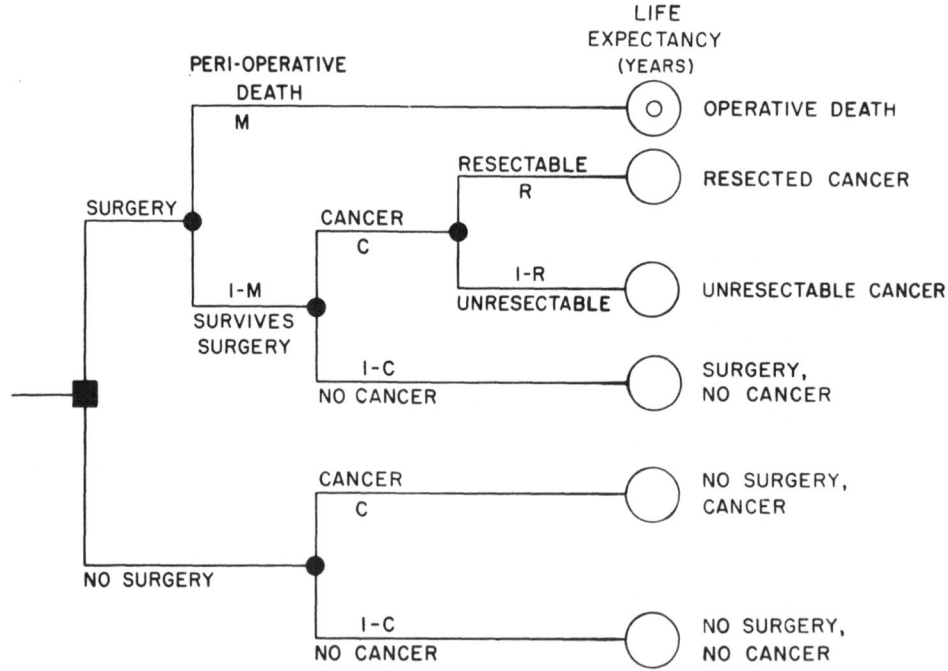

Fig. 1.

This tree has six possible outcomes corresponding to six utilities (the open circles). These utilities are quantitative indices of the relative values of each of the possible outcomes. These are expressed as life expectancy (in years) for each outcome. Thus the worst case, that of peri-operative death, will have a utility value of zero (years). One can also see that the best case, that is the one with the highest utility value, would be no surgery and no cancer.

In order to use this model we need numerical values for all its variables, namely probabilities M, C, and R, and the six utility values. These variables will, in turn, depend on several characteristics (parameters) of a particular patient. The following eight patient parameters are considered: age, sex, smoking history, and the presence and severity of five common diseases (hypertension, angina, diabetes, and history of chronic obstructive lung disease or myocardial infarction). These last five are particularly important in the older patient whose life expectancy may be far more severely limited by them than by a potentially malignant nodule. The probabilities and utilities, for different age, sex, smoking and disease categories, are contained within the programme in tabular form. They have been drawn from the medical literature, from actuarial studies for life insurance companies, and from Canadian and American government statistics. Given these values, it is then possible to calculate the life expectancy for an individual patient both with and without surgery. This is done by sequentially multiplying along the branches of the tree and summing at the nodes ("folding back" the tree). The programme also allows us to consider more general situations by the use of tables and graphics.

In summary, we have developed a decision analysis model and computer programme for the management of patients with a solitary pulmonary nodule. This model may be used for teaching and is a potential aid to the clinician. It will be expanded to include such diagnostic techniques as trans-thoracic fine needle biopsy and trans-bronchial biopsy.

COMPUTER-ASSISTED ANALYSIS OF TRANSTHORACIC IMPEDANCES

R. Duranteau, B. Auvert, B. Dautzenberg, P. Lebeux and
C. Sors
CHU Pitié-Salpétière, 47 Boulevard de l'Hôpital
75651 Paris
Cedex 13, France

Transthoracic impedance at 3 kHz with a well-designed electrode system is a non-invasive method measuring instantaneous changes in left or right pulmonary volumes.[1,2] Our aim was to use a computer to speed up our study on it.

Transthoracic impedance principles: developed independently by Nyboer[1] and Gourgerot,[2] this method uses a four-phase 3 kHz generator delivering 16 VAC. Instantaneous thoracic impedance is related to pulmonary gas and pulmonary blood instantaneous volumes: it falls when blood enters the lungs and rises when air comes into the lungs.[3] Position of the electrodes is most important for experimentation: different lay-outs are used for ventilation study and circulation study. EKG electrodes are used for circulation and self-adhesive silver strip electrodes for ventilation.

Correlation between the pneumotachograph method and the impedance technique is excellent (when changes in pulmonary volumes are considered).

The global signal is the sum of a large pulmonary ventilation signal and a much smaller circulation one (i.e. perfusion of the lungs). To see the latter on an ordinary scope, the patient has to stop breathing, which might be impractical in some cases.

The computerised measuring unit is built up with a dual-channel impedance-to-tension converter, a dual channel A/D converter (12 bits) and of an 8 bit-6502 based microcomputer with 48 KO RAM and two 102 KO discs. Sampling rate is 110/s and the languages used are "basic" and "assembler."

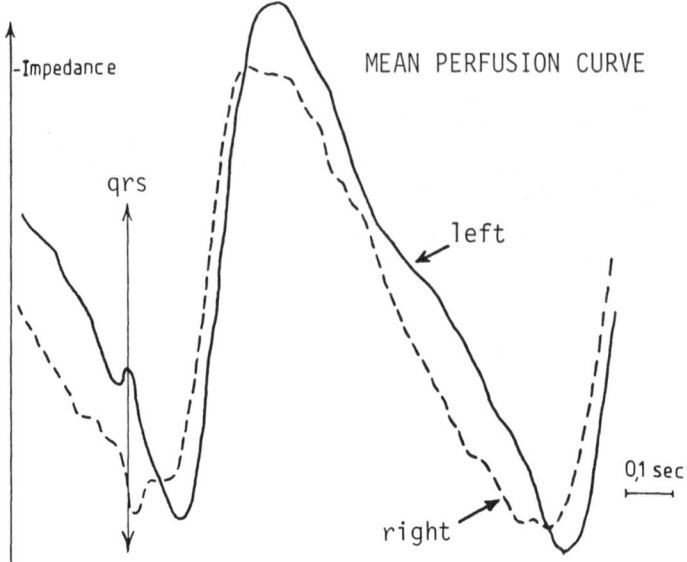

Fig. 1.

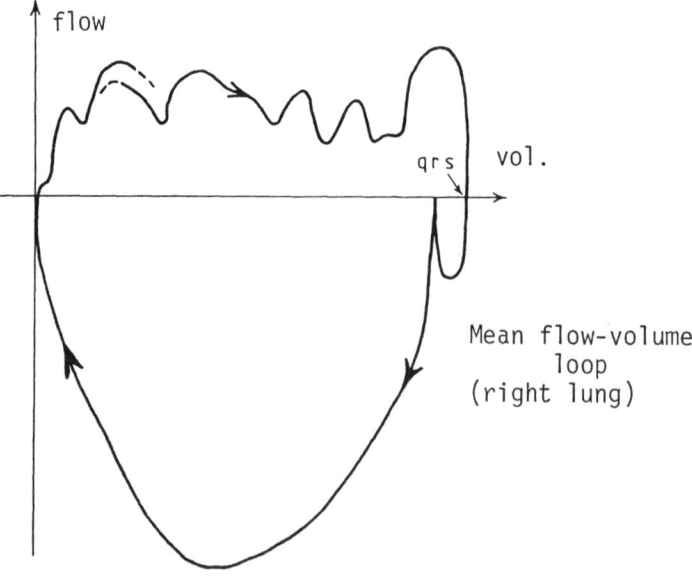

Fig. 2.

Preliminary results:

1. a Fourier-transform of the signal was used for frequency analysis: clear separation of the two "fundamentals" and their related harmonics helped to design our analogical and digital filters.

2. A mean perfusion curve was made by adding 30 intervals of 2s (Fig. 1). It shows a cardiac artifact, then a rise when the systolic pulse reaches the lung, then a decay when blood gets out of the lungs. Several parameters can be measured: QRS to beginning of increase flow-rise, flow-decay, phase-shift between the two curves, etc.

3. A digital derivative gave the left and right flow signals, enabling the display of a flow/volume loop for left or right pulmonary circulation, not ventilation (Fig. 2). Axis are the same as the well-known F/V loops for ventilation. The systolic part of the curve is its lower half.

4. A similar method was implemented in ventilation studies with lowpass digital filters to obtain the ventilation F/V loops for each lung: although early results correlate to the pneumotachographic method, the study is still under development.

IN SUMMARY

1. Computer analysis of physiological signals can be implemented on low-cost microcomputers, allowing a portable yet efficient system.

2. The transthoracic impedance method is an easy-to-use way of studying and checking pulmonary functions. The former has to be developed with a statistically significant number of patients, normal or pathologic ones, to increase our knowledge of the waveforms.

REFERENCES

1. J. Nyoboer, Bilateral pulmonary function by electrical impedance spirometry, Harper Hôsp. Bull. 22:232-45 (1964).

2. L. Gougerot, P. Monzein, J. B. Leblond, B. Gamain, Estimation de la ventilation séparée des deux poumons par la mesure des variations de l'impédance transthoracique in: Acquisitions dans les mesures d'impédance bioélectriques, (Compte-rendu du deuxième Congrès International d'Impédances Bioélectrique). Lyon S.P.C.M., 202-10 (1978).

3. B. Dautzenberg, C. Sors, L. Gougerot, L. Monzain, Measurement of the circulation and ventilation of the left and right lungs by transthoracic impedances: interest for clinical investigation, Am. Rev. Resp. Dis., 119:104 (1979).

PREDICTION OF TISSUE PERFUSION AND HEAT GENERATION RATES BY AN IMPROVED HEAT CLEARANCE TECHNIQUE

R. C. Eberhart, A. B. Elkowitz, A. Shitzer*

Dept. of Surgery, University of Texas Health Science
Center, Dallas, USA
*Dept. of Mechanical Engineering, Technion
Haifa, Israel

Thermodilution (heat clearance) was introduced 50 years ago for prediction of blood flow rate in tissue. The method lost favour because of the size of the temperature probes, technical requirements for generation of a tissue temperature perturbation, and lack of appreciation of interplay of conduction, heat storage, and convective transport. Recent progress has eliminated several of these objections. Tissue temperature shifts may be reliably sensed to $\pm$ 0.05°C with commercially available probes 0.8 mm in diameter. A heat balance in tissue has been developed, called the bio-heat transfer equation, which allows explicit account of all heat transport mechanisms.[1] The biological medium is assumed to have homogeneous and isotropic thermal properties. Tissue blood flow is non-directional at the capillary level, that is, the capillaries are assumed to be randomly oriented with respect to their arteriolar and venular connections. Convective heat exchange occurs only in this capillary system; arteriovenous anastomoses and other arteriovenous heat exchange mechanisms play no role. Tissue blood flow is assumed so low that tissue and end-capillary temperatures are in equilibrium. Tissue blood flow may not vary in the thermal influence region of the probe. Local heat production is ignored, but can be introduced with little increase in mathematical complexity. Under these conditions the one dimensional bio-heat transfer equation in tissue is

$$\underset{\text{storage}}{\rho C \frac{dT}{dt}} = \underset{\text{conduction}}{K \frac{d^2 T}{dx^2}} + \underset{\text{convection}}{W_b C_b (T_b - T)}$$

ρ = density K = thermal conductivity
C = heat capacity   subscript b denotes blood

Tissue temperature, T, varies with time, (t), and position, (x), under the influence of blood inlet temperature, T_b, tissue blood flow. W_b and the initial temperature distribution. If the temperature or heat transfer at the boundaries of the tissue is prescribed, the equation may be solved for the temperature field. Estimation of the perfusion rate may then be accomplished, based on the temperature solution. One measures the temperature field experimentally and compares it with the solution of the bio-heat transfer equation at all times and positions. The perfusion, W_b, is varied until the difference between theoretical and experimental distributions is minimised, in a least squares sense. Once the experimental and theoretical temperature distributions match over the appropriate period, it is concluded that the correct perfusion rate has been determined. Results of this method for perfusion distribution in the left ventricular free wall are shown in Fig. 1.

A number of complex physiological systems have been successfully handled by this, and related techniques. The perturbing effects of large blood vessels in the vicinity of the probe can be detected and the blood flow recovered.[2] Simultaneous multiple measurements are

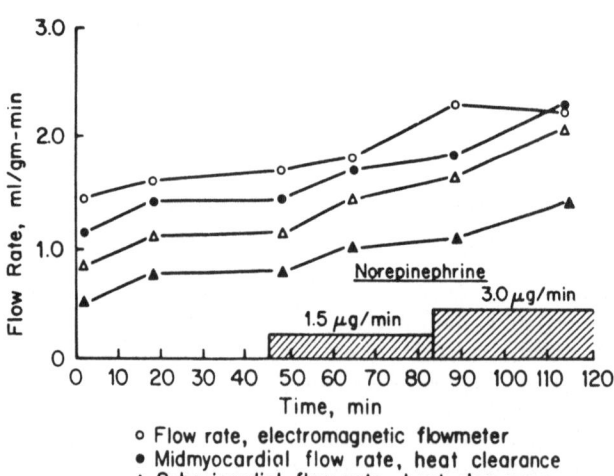

o Flow rate, electromagnetic flowmeter
• Midmyocardial flow rate, heat clearance
△ Subepicardial flow rate, heat clearance
▲ Subendocardial flow rate, heat clearance

Fig. 1.   Measurement of myocardial blood flow in the anterior wall
          of the left venticle in the dog. A pedicle was created for
          control of blood flow, measured by the electromagnetic flow
          meter. The heat clearance technique described in the text
          was used to predict myocardial blood flow at 3, 6 and 9 mm
          depth. Blood flow was varied by intravenous infusion of
          norepinephrine. From Hernandez et al. Amer. J. Physiol.,
          236:H345, 1979.

possible using inexpensive temperature measuring equipment which can be readily implemented in the laboratory of surgical amphitheatre. Time varying, or position varying blood flow can be estimated, but with a substantial loss of accuracy.[3] Engineering sensitivity analysis has outlined the experimental errors most likely to cause error in the predicted tissue blood flow rate. Errors in temperature measurement are more important, by two orders of magnitude, than errors in thermal conductivity, heat generation rate, and other model parameters in distorting the perfusion estimate. Absolute system measurement accuracy of $\pm$ .05°C is necessary to predict perfusion rate to 20% accuracy in a one dimensional perfusion model.

If the assumptions underlying the bio-heat transfer equation are appropriate for the organ bed in question, the bio-heat transfer equation can be used to adequately predict the macroscopic temperature distribution, and solution of the inverse problem, i.e. matching of experimental and theoretical temperature profiles, provides a low cost, semi-continuous means for measuring the perfusion rate.

REFERENCES

1. R. C. Eberhardt, A. Shitzer, E. H. Hernandez, Thermodilution methods: Estimation of tissue blood flow and metabolism, Ann. NY Academy of Sciences, 335:107-131 (1980).
2. H. F. Bowman, T. A. Balasubramaniam, M. Woods, Determination of tissue perfusion from thermal conductivity measurements, Paper 77-WA/HT-40. ASME, United Engineering Center, New York (1977).
3. A. Shitzer, A. B. Elkowitz, R. C. Eberhart, Temperature profiles calculated in tissues subjected to non uniform blood flow distributions. Advances in Biomedical Engineering. ASME, United Engineering Center, New York (1980).

SEMICONDUCTOR ELECTROLYTE ANALYSIS: PROGRESS TOWARDS

INDWELLING CHEMICAL SENSORS

R.C. Eberhart, T.I. Thomasson, R. Sken, K. Wiemer,
G. Cumming, M. Judy and G. Szabo

Department of Surgery
University of Texas Health Science Center
Dallas, Texas

A pH sensitive field effect transistor is a semiconductor device in which the metal gate is removed and the exposed silicon nitride/ oxide insulating layers chemically interact with H^+ ions, thereby modulating the transistor voltage-current relation.[1] An early model of our pH ISFET is depicted in Fig. 1. Further treatment of the insulating layer with an ion selective membrane, such as an ionophore-loaded polymer over the active region allows selective filtration of the ion of interest, e.g. Na^+, K^+, Ca^{++}, which may then also modulate the transistor electrical characteristics.[2] Using such techniques a number of chemical sensors, called ion sensitive field effect transistors, (ISFET) may be created. The number of sensors which may fit on a single chip is only limited by the mechanical placement of the ion selective membrane.

ISFETs are potentially attractive for indwelling blood electrolyte analysis, owing to their small size, economy, robust nature and the ability to combine several sensors and logic designs in a single chip by conventional photomask semiconductor processing techniques. Recent work in ISFET technology has focused on three fundamental problems: the principles of operation of the device, stable operation in electrolyte solutions and the avoidance of thrombogenesis and other artifacts at the sensing surface.

Original investigations of the ISFET indicated ion exchange between the electrolyte and nitride layer dominated the electrochemical process and produced a Nernst response, e.g., voltage shift proportional to log (ion concentration). However, at high electrolyte concentrations and in the absence of appropriate impurity atoms in the nitride layer, space charge effects modify the potential distribution at the interface, and thus change the relationship between

121

Fig. 1.   A two channel pH ISFET. The extended linear regions at the
          right are the pH sensitive gate regions. Conventional metal
          gate FET structures are in series with the ISFET circuits at
          left. Chip size is 0.8 x 2.0 mm.

voltage shift and ionic concentration. This phenomenon must be
included in order to properly describe the behaviour of the ISFET.[3]
Studies of the pH ISFET suggest that H^+ diffuses into the nitride
layer. This modifies the electrical field set up by the electrolyte
solution and acts as a long-term drift source. Account must be taken
of this effect if long-term stability is to be obtained. Electrical
shortening of early ISFETs, due to ineffective polymer or elastomer
based sealants, is expected to be controlled by hermetic sealing
techniques available in advanced semiconductor processing facilities.
A number of blood components can also produce a charge layer at the
ISFET interface: among these are charged and polar molecules in serum
and on cell membranes. In addition to adding artifact to the ISFET
signal, these sources may also lead to thrombogenesis. Methods for
controlling these artifacts have been proposed, based on carbon or
albumin coated microporous membrances. Cell migration to, and
thrombogenesis on the sensing surface would be inhibited, but electrolyte ions can diffuse to the interface. Prospects for ultimate
development of a practical, multiple ion sensing ISFET chip appear
promising.

REFERENCES

1. J. M. Zemel, Ion sensitive field effect transistors and related
       devices, <u>Analyt. Chem.</u>, 47:255A (1975).
2. J. Janata and R. J. Huber, Ion sensitive field effect transistors,
       <u>Ion Select</u>. 1:31 (1979).
3. W. M. Siu and R. S. C. Cobbold, Basic properties of the electro
       electrolyte-SiO_2-Si system: physical and theoretical aspects,
       <u>IEEE Trans. Biomed. Eng.</u> ED26:1805 (1975).

THE SYSTEM FOR QUANTITATING THERMAL-DYE EXTRAVASCULAR LUNG WATER

U. Pfeiffer, M. Birk, G. Aschenbrenner, G. Blümel

Institute of Experimental Surgery of the Technical
University
Munich, FRG

The most common technique for in vivo determination of extra-vascular lung water is the double-indicator dilution method based on the fact that one indicator remains intravascular, whilst the other diffuses from the capillary system.

The double isotope-method, first described by Chinard[1] was disappointing because it measured only 70% of the lung water in cardiogenic oedema and 30% in oncolic oedema.

A new method for detection of pulmonary oedema was the intro-duction of thermodilution. Indocyanine green is used as an intra-vascular marker and heat - and cold - as the diffusible indicator. Extravascular lung water is calculated as the difference of the appropriate mean transit times multiplied by the cardiac output.

EQUIPMENT AND METHODS

In order to reduce sources of mistakes involved in indicator dilution studies, we developed a set up for standardised measure-ments (Fig. 1).

The head of the system is a mini computer for monitoring and control of the peripheral device. It induces the injector to inject 10 ml cold indocyanine green - solution ECG - and airway pressure-triggered. A special fibreoptic-thermistor-catheter senses the two dilution curves in the ascending aorta and feeds them into the processor, which calculates cardiac output, mean tran-sit times, and the intra-(IVV) and extravascular space (EVV).

123

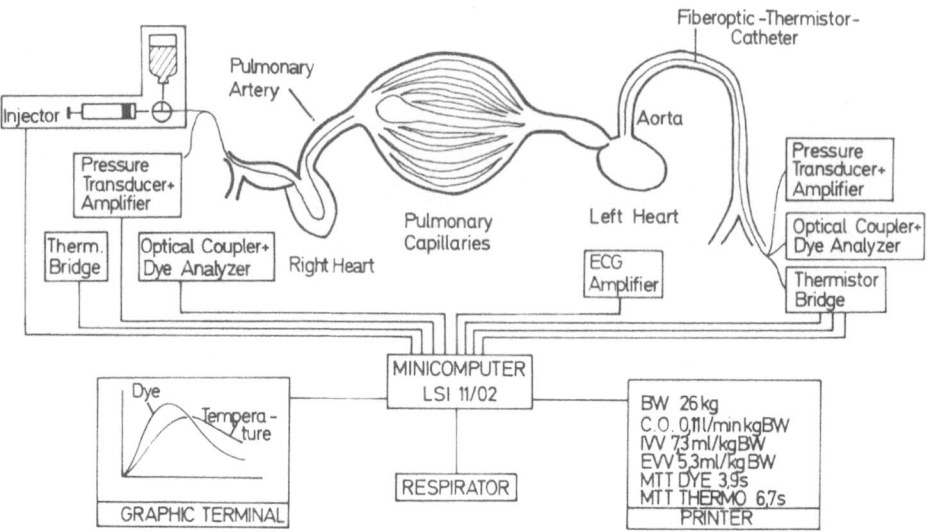

Fig. 1. Monitoring of pulmonary circulation and extravascular water.

    In order to prove the accuracy of the thermo-dye-set up, we
produced pulmonary microembolism in 11 dogs by infusion of 50 mg/kg
BW AMCA for 15 min and 300 NIH thrombin/kg BW for 30 min. Subsequently either 24 ml/kg BW Dextran 40 or the same amount low
molecular weight HES were infused over 6 hours. 15 min before the
amount was sacrificed, 20 µCi ^{51}Cr labelled RBC were injected for
the gravimetric determination of lung water. A last extravascular
lung water measurement was performed - EVV ranged between 10 and
40 ml/kg BW -   then the animals were killed and the lungs removed
in a standardised time.

    The preparation process yielded the weight of intrapulmonary
blood, extravascular lung water (EVLW) and - dry weight (EVDW).

RESULTS AND IMPLICATIONS

    A regression analysis between the last in-vivo measurement of
EVV with our system and the gravimetrically determined extravascular
water content resulted in

    EVV = 3.87 + 1.14 EVLW ml/kg BW, r = 0.98.

The y-intercept of 3.9 ml/kg BW suggests, that the technique also measures some extrapulmonary extravascular space. By introducing a second fibreoptic-thermistor-catheter into the pulmonary artery we observed that a certain amount of the surrounding tissue is detected. This depends upon the surface-contact time relation.

On account of the anatomy of the lung this amount should be the same on either side of the lung during a single circulation time. A new regression analysis between thermo-dye lung water corrected for the extrapulmonary extravascular space yielded a regression line meeting the zero-point

$$(EVV_{corr.} = 0.15 + 1.15 \text{ EVLW ml/kg BW, } r = 0.96).$$

However, the thermo-dye EVV actually represents the caloric equivalent of the extravascular tissue. The connection between EVV and EVLW is given by:

$$EVV = \frac{EVLW \times C_W + EVDW \times C_S}{C_E + C_p}$$

$(C_E; C_p$ = specific heat capacity of RBC and plasma, $C_W; C_S$ = specific heat of water and solid components).

Another regression analysis between thermo-dye EVV and thus synthesized EVV from gravimetric data led to EVV (TD) = 2.46 + 1.07 EVV (G) ml/kg BW, r = 0.98. The remaining slope deviation from the line of identity of 7% can be attributed to errors in the in-vivo as well as in the post-mortem measurements.

We conclude from our results, that maximum standardisation of the measurement procedure is indispendable and can only be achieved by application of a computer controlled system.[3] Thus the system described follows exactly changes in extravascular lung water from clinically not detectable increases to frothy oedema.

REFERENCES

1. F. P. Chinard, T. Enns and M. F. Nolan, Indicator Dilution Studies with "Diffusible" Indicators, Circ. Res. 10:473-490 (1962).
2. U. Pfeiffer, M. Birk and G. Blumel, Ein vollautomatischer Thermodilutionsinjektor. Biomed. Techn. 24, Ergänzungsband: 60-61 (1979).
3. M. Birk, U. Pfeiffer, G. Aschenbrenner and G Blümel, An Automatic System for Measurement of Lung Water Volumes. In: Digest of the Combined Meeting: XII International Conference of Medical and Biological Engineering and V. International Conference on Medical Physics, Part 1: 7.4 Jerusalem (1979).

THE GATED PHONOCARDIOGRAM: A METHOD OF DIGITAL ENHANCEMENT

S. B. Pett, Jr., R. Halsall, E. I. Hoover, W. A. Gay, Jr.,
V. Subramanian

Cornell University Medical Center
New York City

Automated heart sound analysis has been slow to develop for a
number of reasons, two of which are signal quality and physiologic
variability. These difficulties can be readily approached through
current microprocessor technology and the sequential overlay average
(gating) of the phonocardiogram (PCG).

METHOD

Data Acquisition

The procedure initially involved the implementation of a multi-
channel, continuous analog to digital conversion and direct floppy
disc storage system on an eight bit microprocessor (IMSAI VDP80).
Following this, the raw PCG was digitised on line (Hewlett-Packard
21050), with a simultaneous EKG, and stored directly to disc at rates
of 200 samples per second for approximately 100 cardiac cycles. The
software drivers were developed in machine language. Trial record-
ings were made on selected patients in a post cardiac surgical ICU
to test the process in a "worst case" setting.

Signal Enhancement

A system was developed in Extended Basic to construct a normalised amplitude envelope (NAE) by averaging selected cardiac cycles.
The onset of the QRS complexes was used to define the gating interval.
These onsets were determined from the local maximum of the second
derivative of the EKG. A histogram of interval durations was then

127

S. B. PETT ET AL.

presented to allow the operator to reject specific intervals from the
averaging process. The raw PCG was replaced with an amplitude envelope of its rectified signal. The NEA was constructed from the
averages of the corresponding points of the selected gated intervals.
A normalised variance of the envelope (NVE) was concomitantly formed
from the variance rather than the average of the corresponding points.

RESULTS

    The raw PCG was of very poor quality, as expected, due to res-
pirator and chest tube artifact. Following the construction of a
NAE, S_1 and S_2 could clearly be seen and timed. The NVE demonstrated
which portions of the NAE were intermittent. (Fig. 1).

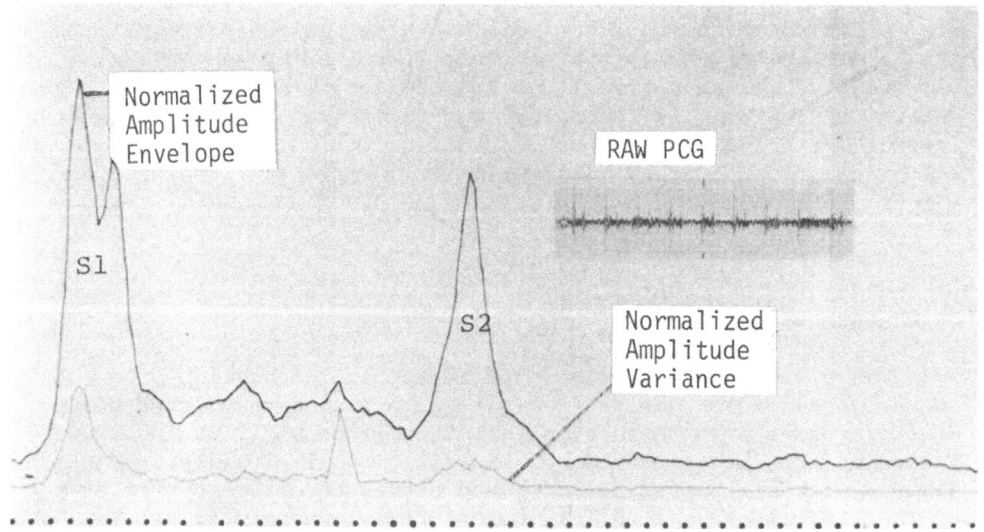

Fig. 1.   The gated phonocardiogram

DISCUSSION

    In general, signal enhancement schemes require the identifica-
tion of some quality which distinguishes the signals desirable from
its undesirable components. A method must then be available to
exploit this difference and extract the desired portion. Frequency
has been one of the most common such qualities, and numerous digital
filtering techniques are available to isolate and attenuate specific
frequency spectra. Gating, on the other hand, requires the

undesirable signal components to occur randomly within some over-riding, identifiable cycle. If sufficient cycles are then averaged point by point, random events will tend to cancel, while consistent events are reinforced. The requirements necessary for application of a gating technique are first: the signal must have some identifiable fundamental cyclicity, and second: multiple cycles must be available for analysis.

There are certain advantages which result from this method of digital enhancement aside from signal clarification. First, the entire interval is reduced to a single cycle. There is obvious economy in storage, but in addition it allows the entire interval to be evaluated as a whole, with decisions being made from a normalised waveform rather than from isolated or selected events. The method is also recursive. There is no theoretical limit to the resolution obtainable in that there is progressive enhancement in the signal to noise ratio with each additional cycle included in the averaging process. Finally, and in many ways the method's most interesting but neglected facet, is that a complementary variance curve (NVE) can be constructed during the averaging phase, which allows one to ascribe a reliability or consistency index to each portion of the cycle.

CONCLUSION

Gating is a useful form of signal enhancement that can be implemented on inexpensive microprocessor based floppy disc systems. It can be applied to any cyclic signal, but is particularly useful in weak or obscure signals such as the PCG. In addition, the analytic promise that the method holds for the PCG may provide a basis for incorporation into future non-invasive patient monitoring systems.

EXACT BTPS-COMPENSATION:

A NEW CONCEPT REALISED WITH A MICROCOMPUTERISED SYSTEM*

D. Rafalsky, H. Sieverts, P. Schwindke

Department of Internal Medicine I and
Helmotz Institute for Biomedical Engineering
RWTH Aachen, D-5100 Aachen
West Germany

For studies of dyspnoea we constructed a new bodyplethysmograph
(BPG) to assess lung mechanics by measuring lung function values as
thoracic gas volume (TGV) and resistance (Raw).

Fig. 1. shows technical details of the equipment. For measuring
during exercise a pressure-constant flow-corrected bodyplethysmograph
is used. This allows the installation of a special air conditioning
system. At the arrow "air in" on the right fresh air passes through
the box pneumotachograph into the chamber. Four ventilators produce
a turbulent airstream. The constant flow through the chamber is
achieved with a laval nozzle. The curved arrow "body position"
demonstrates that a patient can be examined in all body positions
from 0 to 90 degrees. Humidity and temperature probe are situated
in the middle of the chamber fixed over the ergometer.

For measurement the compensation of the chamber signal to the
thoracic gas compression and decompression must be exact. The
chamber signal is influenced by a complex of signals.

Two different procedures are used to correct the chamber signal.

1. Compensation with breathing air prepared in a bag under BTPS
   conditions ($37^{o}C$, water saturated).
2. Breathing ambient air and compensation with electronic BTPS
   correction by the principle of Bargeton.

---

*Supported by Deutsche Forschungsgemeinschaft Proj. No. Me 619.

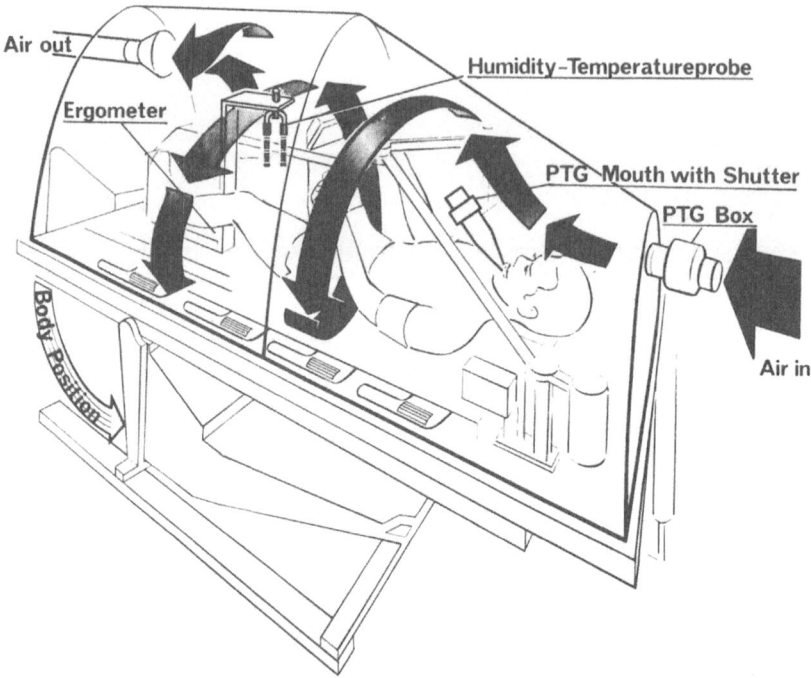

Fig. 1.   Air Circulation and the position of temperature - and
           humidityprobe of our bodyplethysmograph (Open Box Type)

In our experience two facts are important:

a)   The measurement of airway resistance is affected by the
     type of air breathed. The values obtained with the second
     method by Bargeton were, on average, nearly 25% lower than
     those recorded with the first method.

b)   During routine measurements in a bodyplethysmograph we con-
     trolled the climate of the chamber by measuring the humidity
     and temperature. Changes of these climate-factors lead to
     significant errors in the chamber signal. Also the body
     temperature itself changes, if the patient exercises on the
     bicycle-ergometer.

We devised a new method following new conception of continuous
BTPS correction.

THEORY

The airway resistance is defined as the quotient of alveolar pressure P_A and mouth flow V'_M (Formula 1). For the "constant pressure" plethysmograph the alveolar pressure P_A is proportional to the Δvolume body ΔV_B signal

Airway resistance

$$R_{aw} = \frac{P_A}{V'_M} \tag{1}$$

$$P_A = \frac{\Delta V_C - \Delta V_E}{V_L + \Delta V_C - \Delta V_E} \cdot (P_{am} - P_{L_{H_2O}}) \tag{2}$$

Formula 2 describes the alveolar pressure P_A. The compensated V_B signal is substituted by the difference of changing volume in the chamber ΔV_C and also a corection signal ΔV_E. ΔV_C is a measured value of the chamber pneumotachograph. Bargeton evaluated the correction signal ΔV_E empirically by using the breathing of a healthy person. We have tried to calculate the exact correction signal ΔV_E by using the equation for ideal standard gases. The problem is, that the flow measured at the mouth expands inside the lungs. The result is the value ΔV_E. But there is always the same gas flow at both sides.

Exact alveolar pressure

$$P_A(t_1) = \frac{{}_0\int^{t_1}(V'_C - K \cdot V'_M)\ dt}{V_L(t_1) + {}_0\int^{t_1}(V'_C - K \cdot V^1_M)\ dt} \cdot (P_{am} - P_{L_{H_2O}}) \tag{3}$$

with

$$K = \frac{P_{am} - P_{C_{H_2O}}}{P_{am} - P_{L_{H_2O}}} \cdot \frac{T_L}{T_C} - 1$$

Under conditions of ambient pressure, partial pressure of water vapour in the lungs and chamber, temperature of lungs and of chamber, the calculated alveolar pressure for the "constant pressure" body plethysmograph is described in Formula 3. Alveolar pressure P_A is time dependent. The value is the time during an inspiration. The correction signal results from the measured mouth flow multiplied by K. The problem for computing is, that the correct alveolar pressure is influenced by eight values. Five values, which result all from the climate, are linked to the K factor.

ERROR ANALYSIS

$$(4) \quad |\Delta P_A| \stackrel{<}{-} \frac{V_M}{V_L} \cdot (P_{am} - P_{L_{H_2O}}) \cdot |\Delta K|$$

with

$$(5) \quad \Delta K = (K + 1) \cdot \left[\frac{\Delta T_L}{T_L} + \frac{\Delta(P_{am} - P_{L_{H_2O}})}{(P_{am} - P_{L_{H_2O}})} + \frac{\Delta T_C}{T_C} + \frac{\Delta(P_{am} - P_{C_{H_2O}})}{(P_{am} - P_{C_{H_2O}})} \right]$$

With the equation the error can be estimated quantitatively. The error ΔP_A of the alveolar pressure P_A is described (Formula 4). ΔK is determined by the total differential. The error ΔK is proportional to the sum of four relative errors:

temperature of the lungs,
partial pressure of water vapour in the lungs,
temperature chamber,
partial pressure of water vapour in the chamber.

Without measuring these factors the alveolar pressure and the resistance may be in error by as much as 80%. With measuring the environment and computing K the total error of the alveolar pressure and resistance values can be reduced to 10%.

DEVICE CONSTRUCTION

Initially the four probes were linked to the minicomputer PDP 11. Now we use a microprocessor (Fig. 2). A relationship between partial pressure of water vapour and dew point temperature is included in the microprocessor. By using this equation the microprocessor computes the partial pressure of water vapour in the lungs (water vapour saturated) and the partial pressure of water vapour in

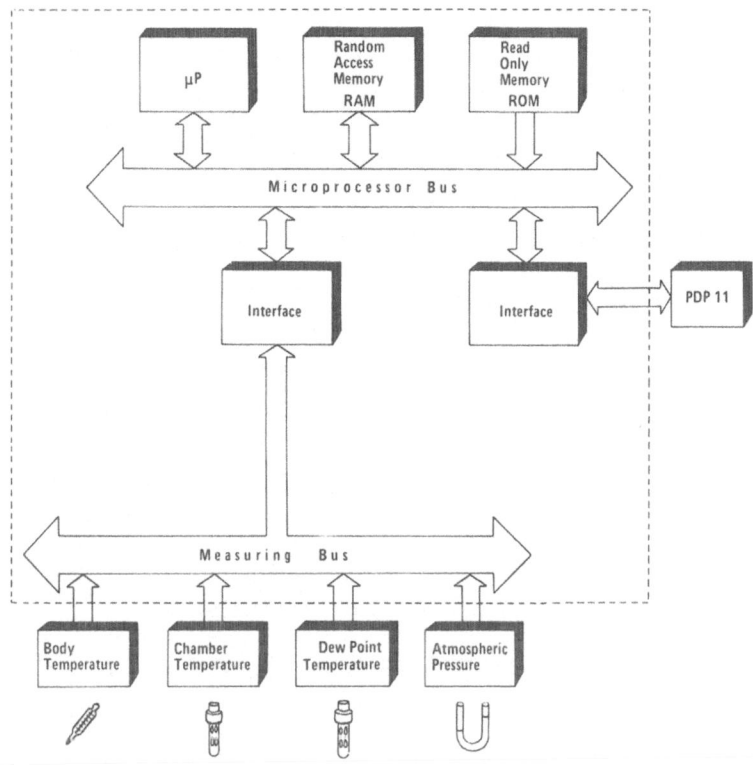

Fig. 2. Design of hardware structure

the chamber by using the measured chamber signal dew point temperatur
ture. With these values and the body probes temperature, chamber
temperature and ambient pressure the microprocessor computes K.

REFERENCES

1. U. Smidt, K. Muysers, W. Buchheim, Electronic compensation in
       differences in temperature and water vapour between in- and
       expired air and other signal handling in body plethysmography.
       Progr. Resp. Res., 4:39-49 (1969).
2. D. Bargeton, G. Barres, R. Lefebure des Noettes, P. Gauge,
       Mesure de la resistance des voies aeriennes de l'homme dans
       la respiration normale, J. Physiol. (Paris), 49:37-40 (1957).
3. J. D. Meyer-Erkelenz, F. Detering, R. Moesges, H. Sieverts,
       Unterschiede des Atemswegswiderstandes bei Atmung mit BTPS-
       Luft und mit Umgebungsluft (elektronische BTPS-Korrektur).
       Prax. Pneumol. 34:455-460 (1980).

RELEVANT DIAGNOSTICAL INFORMATION IN TERMINAL AIRWAY STRUCTURE FROM

COMPUTER ANALYSIS OF EXPIRATORY CONCENTRATION VOLUME DIAGRAMS

Johannes Vogel, Ference Landser,
Gottfried Merker, Burkhard Lachman

Research Institute of Lung Diseases
Department of Pathophysiology
Berlin-Buch, GDR

and

Academisch Ziekenhuis Universiteit te Leuven
Department of Pathophysiology
Leuven, Belgium

Stimulated by the work of Cumming[1] and of Piava[2] and by first experimental results of Smidt and Worth,[3] we tried to look for relevant algorithms, to obtain some new diagnostic information from the shape of expiratory concentration volume diagrams in view of the terminal airway structure.

In patients about to undergo surgery the lungs were ventilated using the Servoventilator ELEMA with constant volume of different levels of constant inspiratory flow. Expiratory concentration volume data are processed by a 12 bit microprocessor unit with fast analog-digital-conversion, to obtain correlations between inspiratory flow and reciprocal $\Delta V/\Delta F$ of the concentration volume diagram in defined concentration levels. At the end of the test parameters of a parabolic exponential term are printed, describing the hypothetical trumpet-shaped cumulative airway structure of the patient: its initial cross section (a), the exponential increase (b) and the effective diffusibility D* of test-gas implying cardiogenic stirring.

The analytical algorithm is derived from model calculations enclosing:

- the normal lung (a = 157, b = 15)
- non-obstructive emphysema and

- general airway obstruction with enlarged respectively
  reduced a, but normal b
- obstructive emphysema with reduced a but enlarged b and
- peripheral airway obstruction with normal a but reduced b.

The interaction between diffusive and convective gas-transport
in terminal airways and the development of a stationary contraction
front during constant-flow inspiration is described by the classical
gas transport equation, which can be changed to a simple differential
equation describing the stationary state. Its numerical solution is
the distribution of the differential quotient dF/dL along airway
length L, which may be integrated and converted to the distribution
of concentration F along airway volume V. Using the working hypothesis, that expiratory front deformation may be negligible, this
distribution is equal to the measurable expiratory concentration
volume diagram in homogeneous lungs.

The type of non-obstructive emphysema gives a flat curve, compared with normal, in contrast to the type of obstructive emphysema,
showing maximal steepness. General obstruction causes increased
steepness in contrast to periferal obstruction. Experimental lowering of inspiratory flow shifts the concentration volume diagram
mouthward to lower expiratory volumes, while its steepness is
increased.

The reciprocal steepness in defined levels is approximated by
a linear function of the flow-diffusion ratio $\dot{V}/D^*$. Its coefficient
of regression is a hyperbolic function of the exponential steepness b
of the model, while the y-intercept is a linear function of its
initial cross section a and of b.

Using this system of approximations there is a theoretically
closed algorithm for calculation of airway parameters from expiratory
concentration volume data (VOGEL).

To go on predicting concentration volume diagrams of inhomo-
geneous lungs we are using a certain lognormal distribution of
volume described by its mean volume-ventilation ratio, that is
FRC/VT, and its inhomogeneity-breath IHB, that is the ratio of the
upper and lower limit on the volume-ventilation-scale concerning
the 90% volume interval. Using the working hypothesis, the low
ventilated space may be a big space and also a low space. It is
possible to derive the distribution of structures and of lung
mechanics, when the mean values b and τ of the total system are
known and the same breadth of inhomogeneity IHB may be accepted.
If the inspiratory flow is distributed in proportion to airway conductance, we obtain the distribution of flow-diffusion-ratio $\dot{V}/D^*$.

For each time t of passive expiration the expired compartment
volume VE_{ti} and its concentration F_{ti} in the local stationary front

may be calculated using the homogeneous approximation system. Expiratory mixing is accepted to occur proportional to the share of compartment flows in total flow.

    Figure 1 shows that pulmonary inhomogeneity shifts the concentration volume diagram to higher expiratory volumes creating an additional dead space volume in comparison to homogeneous lung. This effect increases with increasing IHB.

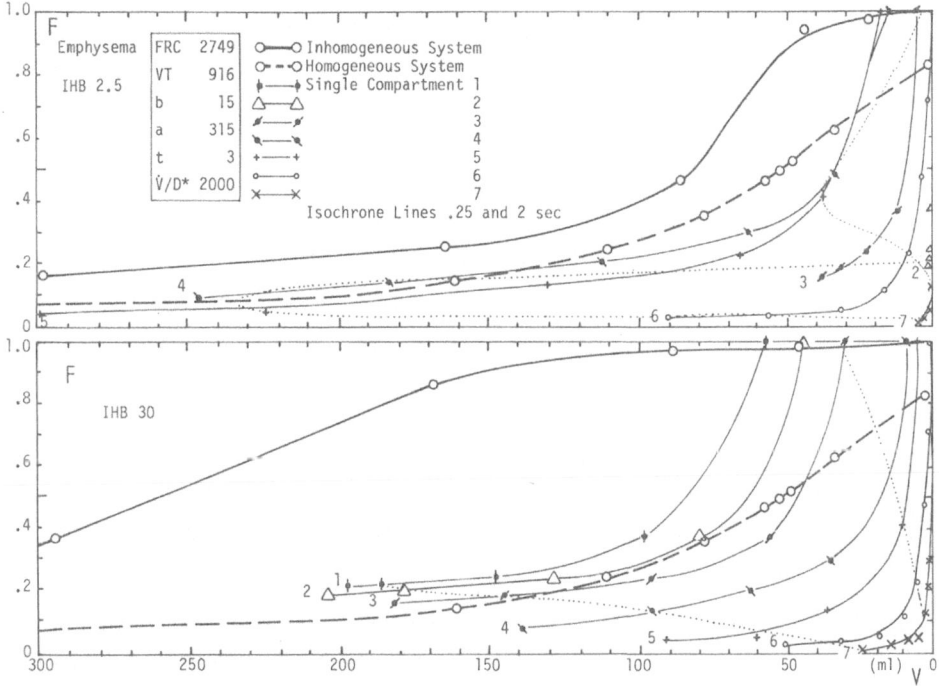

    There may be some evidence to evaluate an analytical procedure for inhomogeneous lungs enclosing classical pulmonary distribution analysis.

REFERENCES

1. G. Cumming, J. Crank, K. Horsfield, I. Parker, Gaseous diffusion in the airways of the human lung, Resp. Physiol. 1:58 (1966).
2. M. Piava, Computation of the boundary conditions for diffusion in the human lung, Computers Biomed. Res. 5:585 (1972).

3. U. Smidt, H. Worth, Diagnostik des Lungenemphysems aus
      expiratorischen CO_2-Partial-druckkurven mit Hilfe eines
      Microprocessors, Biomed. Techn. 22:357 (1977).
4. J. Vogel, G. Merker, E. Mueller, Berechnung der Atemwegsparameter
      aus expiratorischen Konzentrations-Volumen-Daten, Z. Erkrank.
      Atm.-Org. 153:108 (1979).

CHAPTER 6

COMPUTER SYSTEMS IN THE INTENSIVE CARE UNIT

SYSTEM DESIGN FOR THE NEW GENERATION OF INTENSIVE CARE DATA

MANAGEMENT SYSTEMS

T. J. Stafford, A. L. Miller and J. P. Payne

Research Department of Anaesthetics
Royal College of Surgeons in England

At the Royal College of Surgeons we have developed an ITU Data Management System.

In this paper we describe our Systems Design Philosophy rather than the details of the design.

Early in the project we set ourselves certain primary design objectives.

1. very low cost
2. entire data load capability
3. ease of use
4. complete archive
5. reliability

The first objective was that the system should be deliverable at very low cost. We suggest that the stringent economic climate which will probably apply in medicine over the next decade will make the market more cost conscious. In circumstances of increasing price elasticity and competition the demand curve for ITU instrumentation will show more pronounced distinction between the vanishingly small potential market for systems costing more than £40,000 and any low cost system offering similar capability.

Empirical evidence for this proposition is offered by the United Kingdom experience where, since Wythenshaw Hospital acquired a Stansaab System in 1973, there has been a virtual moratorium on buying the high-prices systems on offer. It can be argued that the tight money policies of the NHS presage more general world-wide constraints on medical spending. With this background we set ourselves a target of £3,000-£3,500 for the retail price of hardware for a complete system on a one-off basis.

Our second design objective was based on the premise that any
system worth designing should be capable of handling the entire
data management load. In particular it should be capable of accept-
ing, codifying and displaying all classes of ITU data - not only
routine observations and monitor generated data, but also drugs,
fluid balances and the handling of free format text. Anything less
than this would require a mixed system of computer and manual paper-
work with all the implied complications.

A critical factor in the success or failure of any computer
based system is its acceptability in the user environment - this
means nurses. If it is not acceptable to the ITU nurses on a
routine day-to-day basis it will fail. To meet this end we have
had to examine all potential techniques associated with improving
user friendliness. These include graphics, colour graphics, touch-
sensitive screens, light pens, menus, digitiser pads, function key-
boards, variable function keys, audio signals, warnings, 'Help'
files and pages. Some of these techniques are useful, some are less
so and some are currently expensive. Having examined the functional
utility and forecast the price trends of all of these aids, our
solution represents what we regard as the best compromise between
hardware, expense and user friendliness.

Given the technical capability, as a result of the recent
dramatic fall in the price of bulk storage - in particular that
represented by 8" Winchester technology - it would be inexcusable
to design a system which did not have immediate recall of all relevant patient data, that is the entire stay for all patients over an
extended period of at least one year. In an ITU environment system
failure is unacceptable.

Modern computer technology is inherently extremely reliable at
the chip level but we believe that certain approaches to system
architecture can greatly enhance the reliability of the system as a
whole. Traditional systems architecture as represented by diagrams
A and B (fig. 1) operates an entire multi-bedded ITU from a single
central processor. The bedside data capture point can range from
a simple keyboard to a sophisticated VDU. In the latter case a
heavy communications burden is placed on the central processor
which can lead to 100% increase in system cost. Moreover, any
fault in the central processor means that the whole system is
unavailable until repairs are carried out. Costly standby maintenance thereby becomes mandatory.

Modern concepts of distributed processing are represented by
systems C and D (fig. 1). In an ITU environment this can be
achieved by placing a microcomputer and local storage at each bedside. This approach greatly improves the reliability of the
system since the redundancy involved means that any failure leads
only to a slight degradation in the system performance rather than
catastrophic failure.

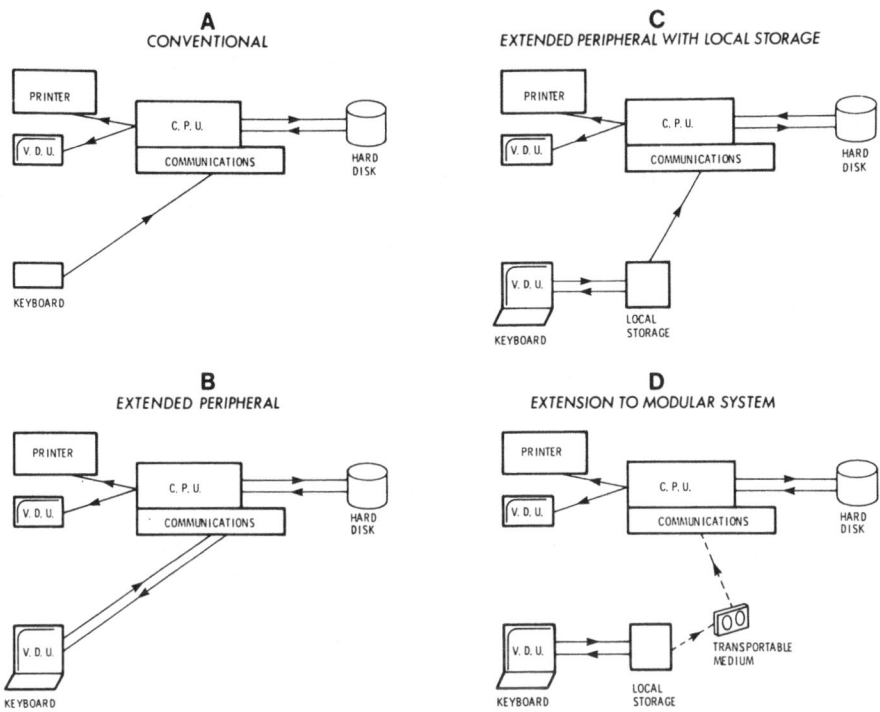

Fig. 1. Approaches to system architecture.

If data is dumped to local storage at frequent intervals, a unit breakdown means that not only the rest of the system (i.e. other beds) is unaffected, but that the absolute minimum of data is lost for the patient in question, and with one back-up unit, total function can be resumed within minutes.

This approach allows complete modularity even at the level of a single bed system. The cost implications are obvious both for purchase and maintenance.

In the course of our project we have examined all the functional components of an ITU Data Management System. In every case we have seen major technological changes and price reductions.

Design in this field is akin to shooting a moving target and therefore throughout this excercise we have been forced to look ahead to the sort of computer systems which will be commercially available at low cost in 1982. When we started this project, systems with our technical requirements, suitable for the bedside, fully packaged, and at our target cost did not exist. With its powerful instruction set and extensive base of user and GEM support we determined that the Z-80 system would be the family from

which these requirements were likely to emerge fastest. Examination
of ITU data processing shows that almost everything is, or can be,
manipulated directly in character form so that 8-bit processors are
entirely adequate and indeed have some advantages over larger
machines. As already emphasized, all the technical components have
shown major price reductions over the last two years and we feel
confident in predicting that packaged systems incorporating all the
necessary capabilities for ITU system will be available for an all-
up cost of between £1,500-£2,000.

   In contrast, the annual cost of a systems programmer capable
of designing or modifying systems of this complexity will have
increased from the present £9,000-£10,000 range to the £13,000-
£15,000 level. Very few ITU units will therefore be able to support
their own systems development. This highlights an area of potential
conflict in that Intensive Therapy Units are by their nature very
individual in their needs. The range of patient care can be broad
or very particular. Individual consultants have their own needs and
preferences. Our approach to this has been to design a system which
at implementation can be modified within defined limits to reflect
local needs. Display page configurations are fixed but the information displayed can be chosen by the consultants responsible.

   In conclusion, the design philosophy for the new generation of
ITU systems must reflect the recent changes both in technology and
economic circumstances and will be completely different from the
approach taken in the last decade. The system which we have
developed at the Royal College of Surgeons has successfully met all
of the objectives that we have listed.

   We gratefully acknowledge the assistance and support of the
Charles Wolfson Trust throughout the whole course of this project.

COMPUTERISED MEDICAL DECISION-MAKING - AN EVALUATION IN ACUTE CARE

R. M. Gardner,* T. P. Clemmer,** A. H. Morris**

*Department of Medical Biophysics and Computing
**Department of Medicine
LDS Hospital/University of Utah
Salt Lake City
Utah, USA***

We have developed methods which utilize the computer to gather and analyse clinical data on all patients in our hospital. If the patient data shows a life threatening result the computer automatically alerts the medical staff.[1] We have evaluated this system and its clinical use over the past two years.

Figure 1 shows a block diagram of the data collection, processing and reporting system used. The HELP computer system[2] consists of a central data based computer system with twelve minicomputers attached. The minicomputers acquire data from areas such as the intensive care units (ICU) (51 beds), clinical laboratory, blood gas laboratory, etc. Once data has entered from any of these sources the data is automatically sent to the HELP computer system. Then the data is automatically stored and processed by the HELP system to determine if the new information by itself or in combination with other data in the patient record (such as another laboratory value or a previous computer-generated decision) leads to a medical decision. The decisions are based on criteria which are stored on magnetic disk and which have been determined from Physician and nurse knowledge, literature searches and from analysis of the computer data base. After the data are stored and processed (usually done in less than two minutes) they become available to the clinical staff by a variety of pathways. For patients in the ICU the results, with decisions, are printed on a printer located in the unit. For the other areas of the hospital, the results are available for review on video terminals. Decisions

***Supported in part by grants HS02463 and GM23095.

147

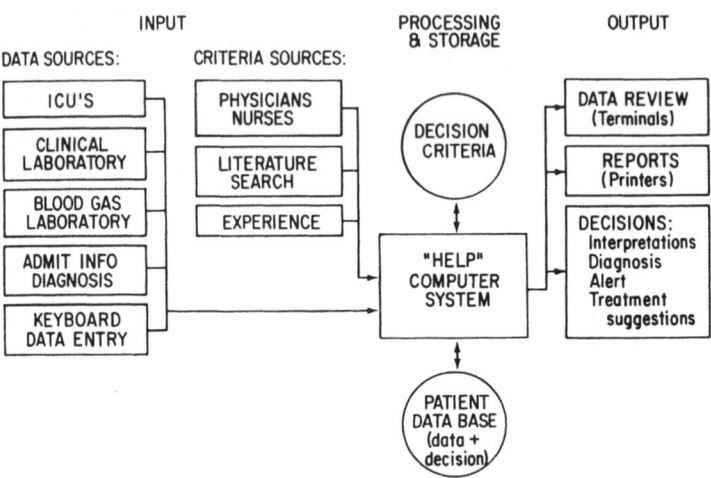

Fig. 1.

made by the HELP system are of the following type: (1) Interpretation,
(2) Diagnoses, (3) Alerts - notifications of life threatening events,
and (4) Treatment suggestions. Most data are automatically gathered
by computers - the system requires little data entry by physician or
nurse, and decisions are automatically generated when new data are
entered or a predetermined time interval passes.

    To assess the effectiveness of the alerting function without
disturbing the convential data communications and review mechanisms
we followed up computer-generated alerts. Nurses followed up every
computer alert but made direct contact with the physician only if the
patient was identified as a study patient and simply noted the type
of care if the patient was a control. The nurse documented the quality of patient care by using several quantitative measures such as
time interval from computer detection of the alert until the first
physician action. They also measured physician compliance to simple
computer generated therapeutic suggestions.

    Based on a two year study we made the following preliminary
conclusions.

    1. Life threatening alerts occurred in about 8% of 15,422 medical
       and surgical cases admitted over a 10 months' period. This

finding is consistent with earlier work we had done with
pharmacy alerts[3] which showed that only 3.0% of patients had
alerts. We found that physician compliance with the pharmacy
alerts was very high (75%).

2. Most alerts fell into one of the following categories.
   A. Major change to patient care - such as metabolic acidosis.
   B. Minor change to patient care - such as haematocrit falling.
   C. No effect on patient care - such as low haematocrit.
3. For those alerts such as metabolic acidosis, where there was
   a major change, we found that
   A. Physician response for the study patients was faster and
      action was taken more often than with the control patients.
   B. The duration of a life threatening alert condition was
      shorter for study patients than for control patients.
   C. The compliance with the computer-generated protocol
      resulted in a more prompt patient recovery.
4. That physicians "learned" quickly to respond to previously
   unrecognized life threatening events. This "learning" con-
   taminated the study and showed the computer alert system to
   have less effect on patient care than it really had.
5. The mortality rate of the "alerted" patients was much higher
   than for the average hospital patient. For example, the
   patients who alerted with metabolic acidosis had an in-
   hospital mortality rate of 18.0%. For the same time interval
   the overall hospital rate was 2.8%.

Physician acceptance of the computer system has been excellent.
Some of the advantages are prompt data communication, data integration, data interpretation, convenient data review, and education
by the computerized data interpretation.

Future challenges include

1. Development of simpler and more efficient methods of entering
   patient information. The "friendly" man-machine interface.
2. Development of computer algorithms (medical care) which are
   action orientated and effective in improving patient care.
   Unfortunately medicine is not quantitative enough to permit
   easy development. Some medical algorithms which currently
   exist are ineffective in improving patient outcome.
3. Methods for assessing the effect and possible benefit of
   computers and other technologies are inadequate, complex and
   expensive.

REFERENCES

1. R. M. Gardner, T. P. Clemmer, Computerized protocols applied to
   acute care, Emergency Medical Services, 7:90-93 (1978).

2. H. R. Warner, Computer-assisted medical decision-making,
    Academic Press, New York (1979).
3. R. K. Hulse, S. J. Clark, J. C. Jackson, H. R. Warner,
    R. M. Gardner, Computerized medication monitoring system,
    Am. J. Hosp. Pharm. 33:1061-1064 (1976).

HIERARCHICAL INTERMITTENT RESPIRATORY TESTING IN THE INTENSIVE

CARE UNIT*

R. M. Peters, J. E. Brimm, S. R. Shackford

Department of Surgery
University of California School of Medicine San Diego
San Diego, California, U.S.A.

and

U.S. Naval Hospital
San Diego, California, U.S.A.

In opening a conference which centres on computers in critical care and pulmonary medicine, it seems appropriate to focus a presentation on the three segments of the name: computers, critical care, and pulmonary. For testing of pulmonary function to be effective in the intensive care unit, it must provide meaningful information within a few minutes of performing a test. Using the available sensors for analysis of pulmonary function, the primary unprocessed signals are of very limited use. Even blood gases need calculations of derived values. Pulmonary testing and monitoring in the intensive care unit have lagged behind cardiovascular monitoring and testing because of this need for signal processing and analysis. For immediate report of results in as simple a measure as spirometry, a computer is essential.

Without a computer, there is no hierarchical test system. All of the tests I shall discuss require some form of computation. In our intensive care units, all the computer processing is done using the Hewlett Packard 5600A system, and all the programmes are available for export to other facilities having the appropriate hardware

---

*Research supported by USPHS Grant Nos. GM 17284, HL 13172, and Office of Naval Research Contract No. N00014-76-C-0282.

and systems software. Level I tests require a pneumotachograph
and a differential pressure manometer to measure airway or trans-
pulmonary pressure. Level II is the above plus an instrument to
measure expired gases (mass spectograph). In addition to analysing
signals from the sensors, programmes interpret the entered arterial
and venous blood gases and use these values to calculate additional
derived values. Cardiac output can also be measured by thermal
dilution, and derived variables calculated.

Since my experience and expertise are with the patient who is
a victim of accidental trauma or the beneficiary of elective surgery, I shall confine this discussion to such patients. Level I
testing measures lung mechanics and work, intermittent mandatory
ventilation (IMV) and spontaneous rate and tidal volume. Level II
testing adds O_2 and CO_2 exchange, end-tidal O_2 and CO_2, deadspace,
functional residual capacity (FRC), continuous respiratory quotient
(RQ), and closing volume.

The end point criteria we used for assessing functional
adequacy are preservation of normal acid-base status, and provision
of adequate available oxygen. Some of the problems in evaluation
of pulmonary function in the surgical patients can be shown by
Level I studies performed on the first, second, and third post-
operative days, the last hospital day, and six weeks to three
months post-operative in adult patients undergoing mid-line
sternotomy and cardiopulmonary bypass for coronary artery bypass
and/or valve replacement. The 55 patients in this study were all
ventilated post-operatively until the morning after surgery with an
intermittent mandatory ventilation system (IMV), with volumes in
the range of 13 to 17 ml/kg. The rate was set at the start to maintain $PaCO_2$ near 40 mm Hg, and fraction of inspired air (F_IO_2) and
continuous positive airway pressure (CPAP) were adjusted to keep
PaO_2 between 85 and 110 mm Hg. CPAP was infrequently used and was
rarely set above 10 cm H_2O. Fifty of these patients were successfully weaned from ventilation in less than 24 hours; five required
continued ventilation. Figure 1 depicts the mean values of these
studies (Fig. 1). In all of these patients, there is a marked fall
in tidal volume (TV), vital capacity (VC), and FRC. First second
expired volume (EV_1) was decreased, and there were also large
decrements in maximum inspiratory pressure (MIP) and maximum expiratory pressure (MEP). Most of these measured volumes returned to
pre-operative levels only on the studies done in the late post-
operative period. These findings in open heart patients are similar to those described by Beecher[1] 30 years ago for laparotomy
patients.

Post-operative spirometry does not predict failure to wean on
the first post-operative day, the point of decision making. The
pre-operative percent of predicted forced vital capacity (FVC) and
EV_1 was significantly lower and FRC higher in the failure patients.

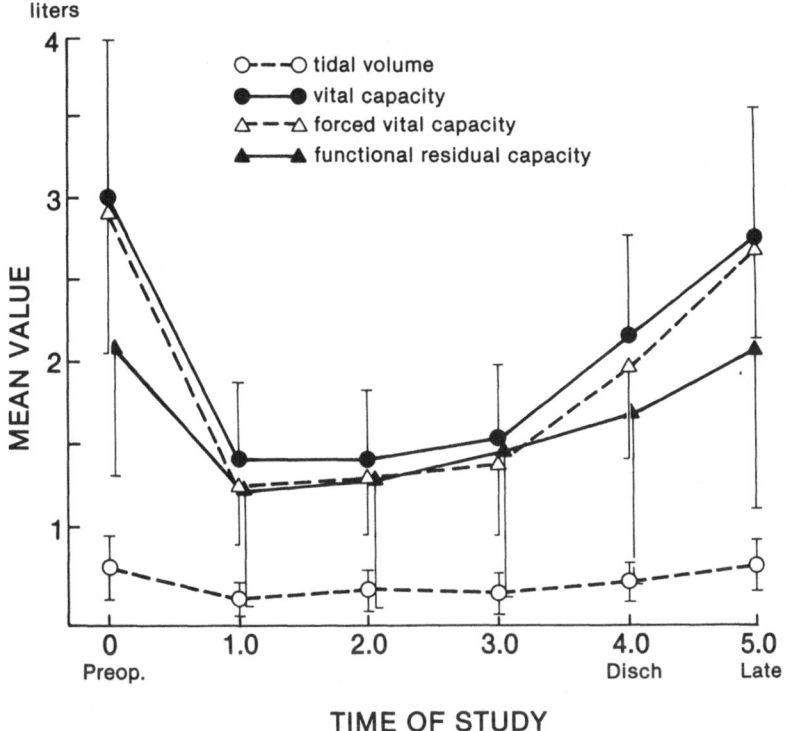

Fig. 1. These graphs depict the fall in tidal volume, vital
         capacity, forced vital capacity, and functional capacity
         in 55 adult patients undergoing cardiopulmonary bypass in
         preop, postop days 1, 2, and 3, discharge, and 6 weeks to
         3 months after discharge.

Hilberman and Osborn[2] found pre-operative studies most predictive
in a similar study.

    These changes in lung volumes are due to dysfunction of the
chest bellows, not the lungs. Splinting due to pain decreases excur-
sion of the chest cage. In the post-operative period, the decreases
in lung volumes due to chest wall disruption are so great that they
obscure post-operative lung dysfunction. Pre-operative studies are
useful predictors of patients likely to have respiratory problems
because they pick out those patients with abnormal lungs. The combination of abnormal lungs and chest cage nearly always requires
ventilatory support for adequate gas exchange.

    For the post-surgical patient, there are useful pulmonary func-
tion tests for making decisions about priorities in the control of
ventilatory therapy.[3] The gold standard for measurement of adequacy

of alveolar ventilation is determined from the acid-base status of
the patient and the $PaCO_2$. We would like a non-invasive test to
determine if alveolar ventilation is adequate to maintain $PaCO_2$ in
the desired range. This would avoid the need for repeated arterial
blood sampling to measure blood gases. End-tidal CO_2 ($PetCO_2$) has
been advocated as a measure of $PaCO_2$ in the immediate post-surgical
and trauma patients. Dr. Brimm will discuss the shortcomings of
$PetCO_2$ later in the conference.

An absolute level of $PaCO_2$ is not the criterion of adequate
alveolar ventilation. A normal $PaCO_2$ of 40 mm H_g in the patient
with a low PaH represents severe deficiency of function - a lack of
ability to hyperventilate to correct the low PaH. If the $PaCO_2$ is
appropriate to the PaH, the patient's ability to sustain the
required level of alveolar ventilation needs to be assessed: the
work required to ventilate the lungs, the ventilation-perfusion
($V_{A/Q}$) coordination of the lungs, the efficiency of the chest bellows
the level of ventilatory muscle conditioning.

For the patient on controlled ventilation, there is no inter-
mediate state between total support and no support. Therefore, the
final test is the absolute stress test - to remove the patient and
see if he makes it. In our experience, an indication of the likelihood of weaning success in the patient on controlled ventilation is
the amount of work required to ventilate the patient. At extremes,
measurement of work is useful. Very high levels indicate that a
ventilator must be used. In the patient requiring long-term respirator support, evidence of decrease in work done on the chest and
lungs by the ventilator is indicative of the patient's ability to
begin ventilating himself.

One of the major advances in ventilator therapy, intermittent
mandatory ventilation (IMV) and continuous positive airway pressure
(CPAP), has lowered the incidence and duration of ventilator support.
IMV lessens the amount of deconditioning of the respiratory muscles,
permitting the patients to assume as much of the ventilatory work
as they are capab e of doing.

The appropriateness of $PaCO_2$ to PaH is the absolute criterion
for adjusting IMV rate and volume. The ratio between ventilatory
work done by the patient and supplemental ventilatory work required
by IMV is a strong predictor of the patient's ability to tolerate
the change in IMV rate. If the measured spontaneous rate rises and
tidal volume falls significantly, the patient will not maintain
adequate alveolar ventilation, and $PaCO_2$ will rise.

We have just completed a study of patients with crushed chest
that illustrates the usefulness of measuring changes in spontaneous
tidal volume and rate.[4] In these patients, work is increased due
to flail chest, and the ability to do this work is limited by the

splinting due to pain. As a result, secretions are retained, FRC falls, and the lung becomes stiff.

Regional analgesia by epidural block allowed us to keep such patients off the ventilator. Effective pain control prevents splinting, and permits effective deep breathing, thus preserving lung-function and avoiding the intolerable combination of sick lungs and deranged chest bellows. On admission, the patients in this study had clinical appraisal of dyspnea and tachypnea. Patients with PaO_2 60 mm Hg, $PaCO_2$ 50 mm Hg, shunt fraction (Q_s/Q_t) 25, or those requiring an operative procedure, were intubated and placed on CPAP and IMV. The rest of the patients received epidural blocks as soon as possible after admission, and these were repeated as needed. Both in patients not on IMV and those receiving IMV and CPAP, a rise in rate and fall in tidal volume and VC were the most useful criteria for repeating epidural injection and controlling IMV rate. Using this treatment protocol, less than 20% of patients required prolonged ventilation, and the duration of ventilation for most of the others was less than 10 days.

Even with adequate alveolar ventilation, PaO_2 may be low. It is not adequate to assess only the PaO_2; it is essential to optimise the oxygen available to the tissues, the product of cardiac output and arterial oxygen content. Surgical patients and those with post-traumatic acute respiratory distress syndrome (ARDS) have hypoxaemia without hypercapnia. This hypoxaemia is due to alveolar collapse. With ARDS, the problem is high permeability pulmonary oedema, which causes a progressive fall in compliance and FRC, and increase in respiratory work and intrapulmonary shunt. The most useful tests for assessing the pulmonary contribution to the hypoxaemia are measurement of Q_s/Q_t and FRC.

To assess the level of testing needed for patients with depressed PaO_2, it is necessary to measure arterial blood gases and calculate the intrapulmonary shunt. The intrapulmonary shunt can be calculated by measuring PaO_2 and F_IO_2 and the alveolar air equation to calculate alveolar PO_2, equations such as those of Kelman to calculate the arterial and pulmonary capillary oxygen contents and assuming an arteriovenous oxygen difference $(A-V\ DO_2)$ of 4.8 volumes per cent. We have shown that if calculated shunt is less than 0.2, assumed A-V DO_2 of 4.8 is safe to use.[5] If the calculated shunt is larger than 0.2, it may be inaccurate. A Swan Ganz catheter must be used to permit measurement of venous oxygen content, pulmonary capillary wedge pressure, cardiac output, and available oxygen. These are the minimal tests necessary for controlling therapy in patients with shunt greater than 0.2.

The importance of measurement of cardiac output and venous oxygen content as part of the pulmonary function testing needs emphasis. Lowering venous oxygen at the same shunt fraction lowers

PaO_2. The most common cause of low venous O_2 content is depressed
cardiac output; thus, depressed cardiac output is a major cause of
hypoxaemia in the patient with an intrapulmonary shunt. Since
excessive CPAP can compress alveolar capillaries and lower cardiac
output, one needs to know whether the increase in CPAP is opening
alveoli or merely collapsing alveolar capillaries and lowering
cardiac output. This can best be answered by measuring FRC. If
cardiac output falls and FRC remains the same, CPAP should be
lowered, since it is not opening new alveoli but is rather over-
distending already open alveoli and lowering cardiac output. If
cardiac output falls but FRC increases, CPAP is performing its purpose of opening alveoli, but it will result in a fall in available
oxygen due to depression of cardiac output. The patient's vascular
volume must be increased to raise left heart filling pressure above
alveolar pressure and maintain cardiac output.

If only Level I testing is available and FRC cannot be
measured, Suter[6] has shown the usefulness of measuring compliance.
To adjust CPAP, we are interested in compliance of only the lungs,
not the chest cage. By using the esophageal balloon to measure
compliance, only the effect on the lungs is measured, and the
patient's muscular effort to breathe is eliminated. If compliance
increases as CPAP is incremented, then new portions of the lung are
being ventilated. If compliance does not increase after raising
CPAP, lung is being overdistended and the incremented of CPAP is
not helping.

There is another Level I signal that has been under utilised -
airway pressure. Most repirators have a slow responding and inaccurate aneroid manometer to measure peak and end-expiratory pressure.
These values need to be accurately measured and sampled for a series
of breaths. The addition of mean pressure gives the pressure that
determines resistance at the pulmonary capillary, the controller of
left heart filling. At constant level of CPAP and constant IMV
tidal volume, monitoring the trend of mean airway pressure can indicate a fall or rise in compliance. A rapid rise of mean airway
pressure, fall in compliance, is the first and a reliable indicator
of a pneumothorax. A slow fall in mean airway pressure signals
rise in compliance and indicates need to lower CPAP.

Automated analysis of changes in airway pressure would represent the lowest level of front end signal device and the simplest
form of computational algorithm. Even this level is still not
available. For the future, we must provide proper computational
interpretation of these simple signals and develop more complex
analyses of higher level signals to enhance the usefulness of the
expensive instruments being introduced to intensive care units.

I have tried to present to you our ideas concerning the use of
pulmonary function in the intensive care unit. We must ask why

these tests have gained such limited acceptance. Let me remind you
that when we first introduced bedside electronic monitors for
systemic pressure and ECG and file analysis of blood gases, many
skeptics scoffed at their usefulness. Only as physicians learned
to use and, unfortunately, at times misuse them, were these new
forms of information accepted.

REFERENCES

1. H. K. Beecher, The measured effect of laparotomy on the respira-
    tion, J. Clin. Invest. 12:639 (1933).
2. M. Hilberman, B. Kamm, M. Tarter and J. J. Osborn, An evaluation
    of computer-based patient monitoring at Pacific Medical
    Center, Comp. Biomed. Res. 8:447 (1971).
3. R. M. Peters, Life saving measures in acute respiratory distress
    syndrome. Am. J. Surgery, 138:368 (1979).
4. S. R. Schackford, R. W. Virgilio and R. M. Peters, Selective
    use of ventilator therapy in flail chest injury, J. Thorac.
    Cardiovasc. Surg., in press.
5. A. R. Shapiro, R. W. Virgilio and R. M. Peters, Interpretation
    of alveolar-arterial oxygen tension difference, Surg.
    Gynecol. Obstet., 144:547 (1977).
6. P. M. Suter, H. B. Fairley and M. D. Isenberg, Optimum end-
    expiratory airway pressure in patients with acute pulmonary
    failure, New Engl. J. Med., 292:284 (1975).

NEW POSSIBILITIES IN PAEDIATRIC INTENSIVE CARE*

R. Frey, W. Bleicher, G. Budwig, E. Hiesinger,
B. E. Kemter, J. Apitz

Zentrum für Pädiatrie und Frauenheilkunde der Universität
Tübingen and
Lehrstuhl und Abteilung für Pädiatr, Kardiologie Institut
für Biomedizinische Technik Stuttgart

Our aims in a computerized intensive monitoring system are:
control of heart and circulatory functions with continuous monitoring
and graphical presentation of vital parameters, which should be
measured non-invasively. By means of graphical presentation of the
parameters in the same picture, it is possible to judge the patient's
state better than with numerical presentation.

We use a modified computerised patient-data management system
with conventional and self-developed bedside monitors. With this
system we have examined 200 children.

At the bedside we have the possibility of monitoring non-
invasively 14 parameters: heart rate, temperature, the transcutane-
ously measured PO_2 ($TCPO_2$), inspired oxygen concentration, the
arterial pressure (SYST, DIAS, MEAN) and six parameters from the
impedance-cardiogram (see below). These parameters are sampled and
stored by the computer every 30 seconds. The feedback of the stored
data and the graphical presentation is indicated on a video screen
near the bedside and by a self-developed reporting system automatically plotted as an intensive care chart.

This chart includes all measured parameters in graphical form
on a sheet (30 x 21 cm). Time-related alphanumerics as lab data,

---

*"Mit Unterstützung durch das Bundesministerium für Forschung und
Technologie Bonn, DVM 161 und DVM 163."

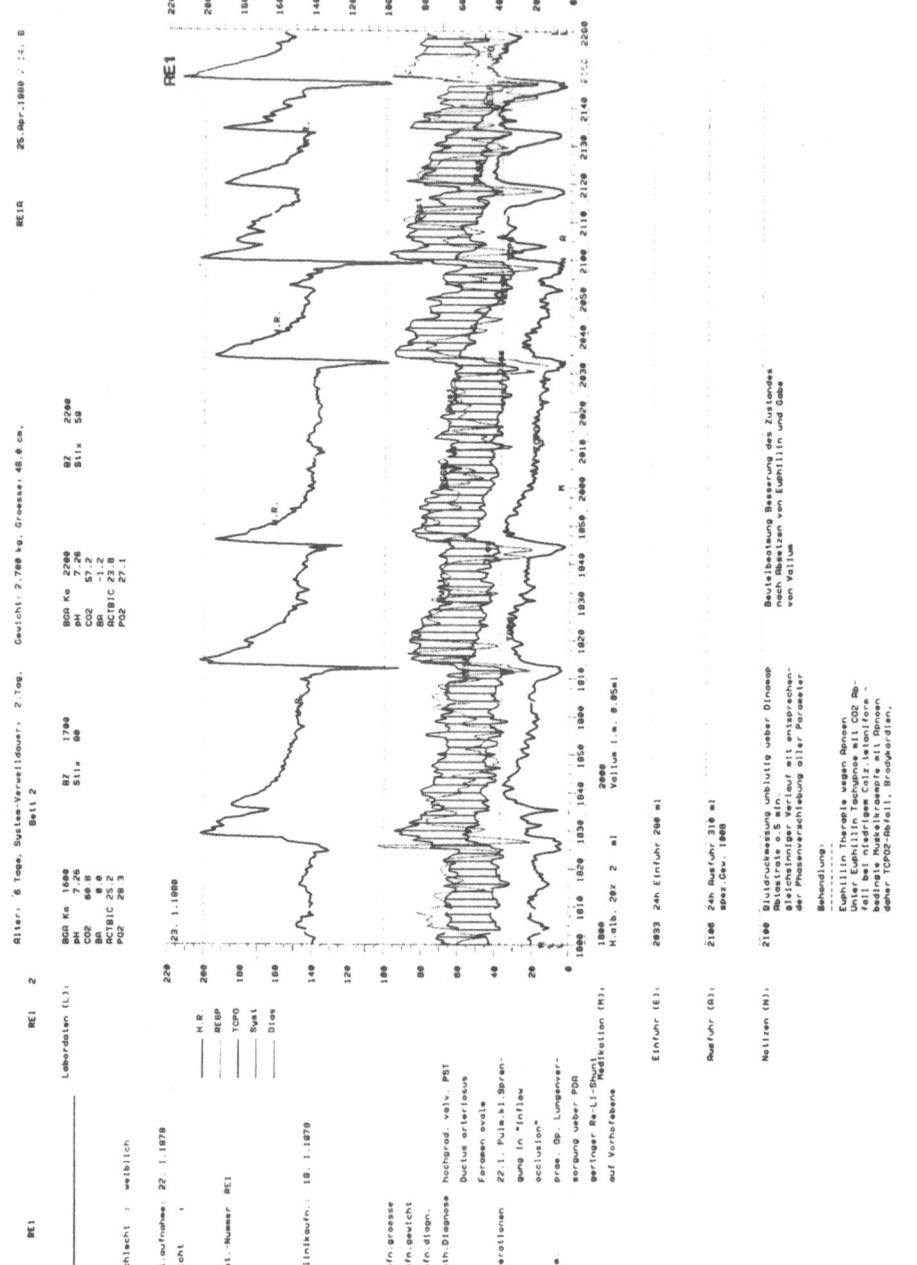

Fig. 1.

medications and other notes are printed chronologically into the same sheet.

The graphical display of heart rate permits statements about its silent course in addition to the control of the effect of drugs on arrhythmias. The continuous monitoring of the $TCPO_2$ and the presentation of its course together with heart and respiration rate results in better control, at which decelerations of the $TCPO_2$ can give earlier warnings. In patients on respirator we obtained better regulation.

Blood pressure can be measured now non-invasively and nearly continuously by means of the DINAMAP system according to the oscillometric principle. The minimal measuring interval is one minute.

The impedance cardiography is a non-invasive method to examine heart function. Changes in blood distribution in the thorax during the cardiac cycle generate impedance changes from which different circulatory parameters can be computed. We developed an automatic device, which continuously calculates from the impedance cardiogram and the ECG beat by beat the following parameters: pre-ejection period (PEP), left-ventricular ejection time (LVET), PEP/LVET-ratio, changes in thoracic fluid balance by changes of the base impedance (Z_0), relative stroke volume and cardiac output.

An example of early diagnosis is given in Fig. 1. Newborn babies with pulmonary atresia and patent ductus arteriosus have typical patterns especially of $TCPO_2$, which indicate the likelihood of the threatening closure of the ductus. We find periodic undulations (6-7 cycles/hour). There are phase-synchronised variations in heart rate, blood pressure and in auscultatory findings.

A PROGRAMME FOR CRITICAL CARE VENTILATOR MANAGEMENT

B. J. Rubin, J. Kunz, L. Fagon, R. Wilson, J. Osborn

Institute of Biomedical Engineering Sciences
The Institutes of Medical Sciences
San Francisco, California 94115, USA

Computer-based critical care monitoring systems provide a continuous display of cardiopulmonary data.[1] The Ventilator Management Programme (VM) uses these data to identify trends in the patient's clinical course and to make specific therapeutic suggestions.

VM is designed for use in the management of post-operative cardiovascular patients (e.g. coronary artery bypass graft). The overall goal is to aid the clinician in weaning patients from mechanical ventilation and discontinuing endotracheal intubation.

INTERACTION

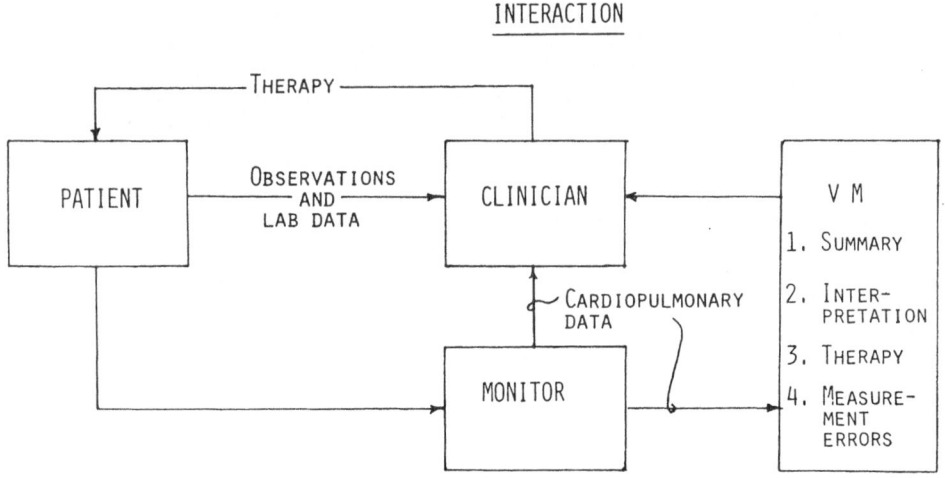

Fig. 1.   Interaction between the patient, clinician, monitor and VM.

Based on the received monitored data, VM (Fig. 1):

1. provides a current summary of the patient's cardiopulmonary status,
2. gives a physiological interpretation of the data,
3. suggests appropriate therapy,
4. detects possible measurement errors.

The programme uses a set of rules which are based on statements of a group of expert clinical practitioners of critical care medicine. Each rule is of the form: if statements about measurements and previous conclusions are true then make a conclusion. Use of rules for representation and manipulation of knowledge is taken from the subdiscipline of Artifical Intelligence called "Knowledge Engineering."[2]

The rule set divided into different sections, each serving a unique function:

Transition determines the mode of ventilation. Since no ventilator presently provides direct computer sensing its operating mode, the ventilatory mode is inferred by analysing characteristic changes in the times of inspiration and expiration (e.g. I/E ratio), the respiratory rate and the maximum inspiratory pressure.

Initialising defines expected values for measured parameters based upon the ventilatory mode. These values are then converted to symbolic ranges such as "ideal," "acceptable," "high" and "low." Thus, limits of acceptability for different parameters vary depending on the clinical context (i.e. ventilatory mode). For example, when a patient is advanced to the T-piece, the upper limit of acceptability for arterial pCO_2 (P_ACO_2) is raised because alveolar ventilation normally decreases temporarily when the mechanical ventilator is disconnected.

Status compares the monitored data and the expected values to make statements about the cardiopulmonary status of the patient. In addition, many status rules incorporate time-dependent trends in measured parameters (Table 1).

Therapy suggests therapeutic interventions for abnormal states, provides a differential diagnosis for particular clinical states (e.g. hypoxaemia indicates the need to repeat a laboratory value) (e.g. follow-up arterial blood gas when hypoxaemia has recently occurred, and defines readiness for the next ventilatory mode in the weaning process) (Table 2).

Instrument detects possible measurement errors and recognises interventions which may cause spurious data (e.g. suctioning).

Table 1. Example of Status Rule

"IF" conditions depend on static measures as well as time-dependent trends.

Status Rule

Status Rule: Stable Hemodynamics

Applies to: All

If: SYS is acceptable
                MAP is acceptable
                HR is acceptable
                MAP does not change by 15 mm Hg in 15 minutes
                HR does not change by 20 beats in 15 minutes

Then: The hemodynamics are stable

Table 2. Example of Therapy Rule

The criteria for beginning a T-piece trial are listed

Therapy Rule: Readiness for Weaning

Therapy Rule: Assist to T-piece

Applies to: Assist

If: Hemodynamics are stable
                Hypoventilation not present
                RR is acceptable
                Oxygenation is acceptable
                Patient in assist for greater than 30 minutes

Then: Consider placing patient on T-piece if patient
                awake and heart rhythm is stable

DISCUSSION

The objective of the programme is to aid the clinician in managing the vast amounts of information generated in an intensive care unit. VM accomplishes this goal by providing a data summary which describes trends in the patient's long-term course and alerting the user to potentially harmful acute changes. VM interprets this data and then makes specific therapeutic suggestions.

It is expected that the programme will be more consistent in its data summary, interpretation and decision making than a group of physicians evaluating the same clinical situation. Of course, therapeutic decisions ultimately depend upon the clinician's bedside impressions. Indeed, the success of this type of programme is dependent upon a "friendly" physician - computer interaction.

REFERENCES

1. J. J. Osborn et al., Computation for quantitative on-line measurement in an intensive care ward, in: "Computers in Biomedical Research," New York, Academic Press (1969).
2. P. H. Winston, Artificial intelligence, Reading MA, Addison-Wesley (1977).

A MONITORING AND DATA ACQUISITION PROGRAMME FOR SURGICAL

INTENSIVE CARE

E. Monoz, A. Sims, J. Dawson, D. Dove, W. Stahl,
L. R. M. del Guercio

Department of Surgery, New York Medical College
Metropolitan Hospital Center, Shock Trauma Unit
New York 10029, USA

Physiological and biochemical data from patients in the Shock
Trauma Unit at the Metropolitan Hospital Center are processed,
stored, analysed and manipulated in a low cost system using extended
"basic" language.

Our equipment consists of an 80 80 S100 BUS based microcomputer
system manufactured by processor-technology, 2 - Thinker Toy 8"
double density disc drives, a Hewlett-Packard graphics plotter model
7221-A, a qume-sprint impact printer and 12" monitor through which
prompt verification of data entry is performed. Approximate cost:
$10,000.

Under CP/M control the APP (Automated Physiologic Profile) programme developed by Cohn, Del Guercio and colleagues is implemented
in micro soft "basic" and booted into memory on start up.

Video prompts request data from inexperienced operators which
are then presented in block form for correction of re-entry of data.
Selection of plot type and pen colour is under soft-ware control.
Automatic plotting in bar graph form on preprinted sheets is performed on the Hewlett-Packard flat-bed plotter.

After the computer has performed numerical analysis, derived
data is then presented to the operator on the monitor. The operator
may then choose file storage on floppy disc. Using algorithms derived from standard physiologic equations, derived data such as C.I.,
L.V.S.W. are calculated. Graphic hard copy is produced under micro
processor control.

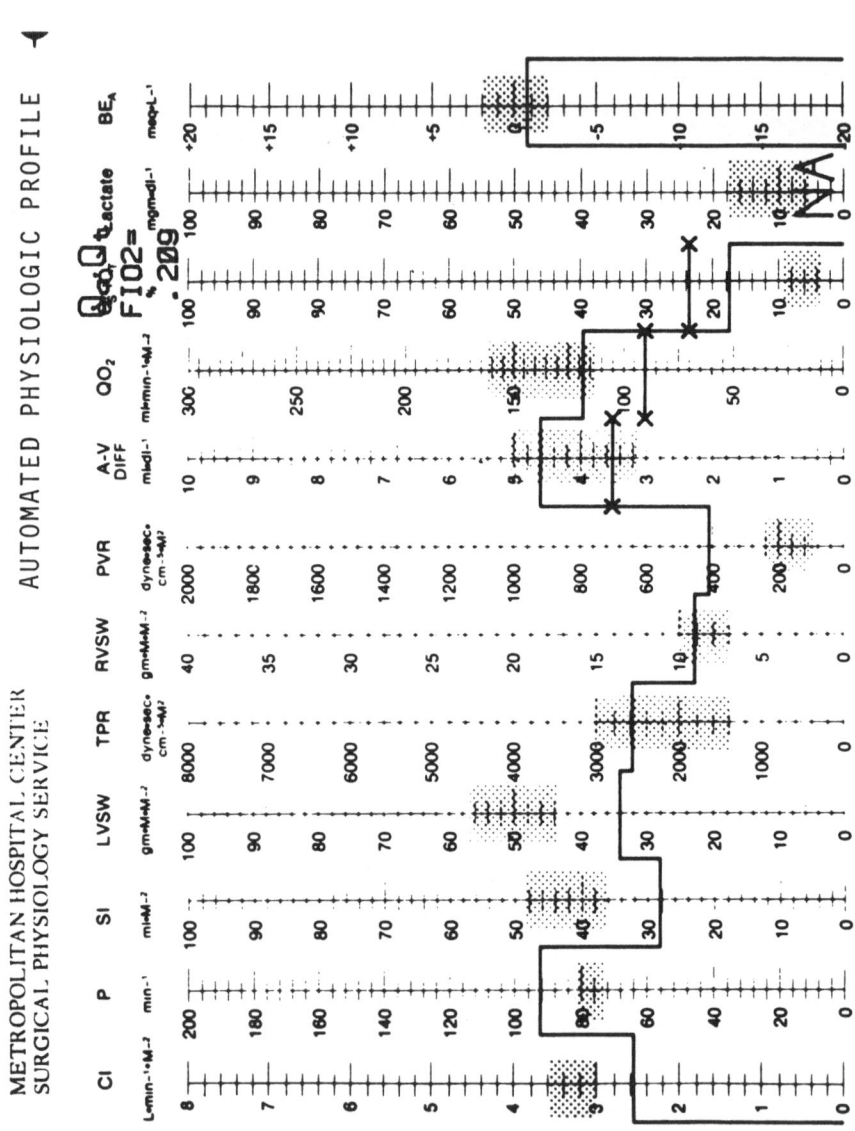

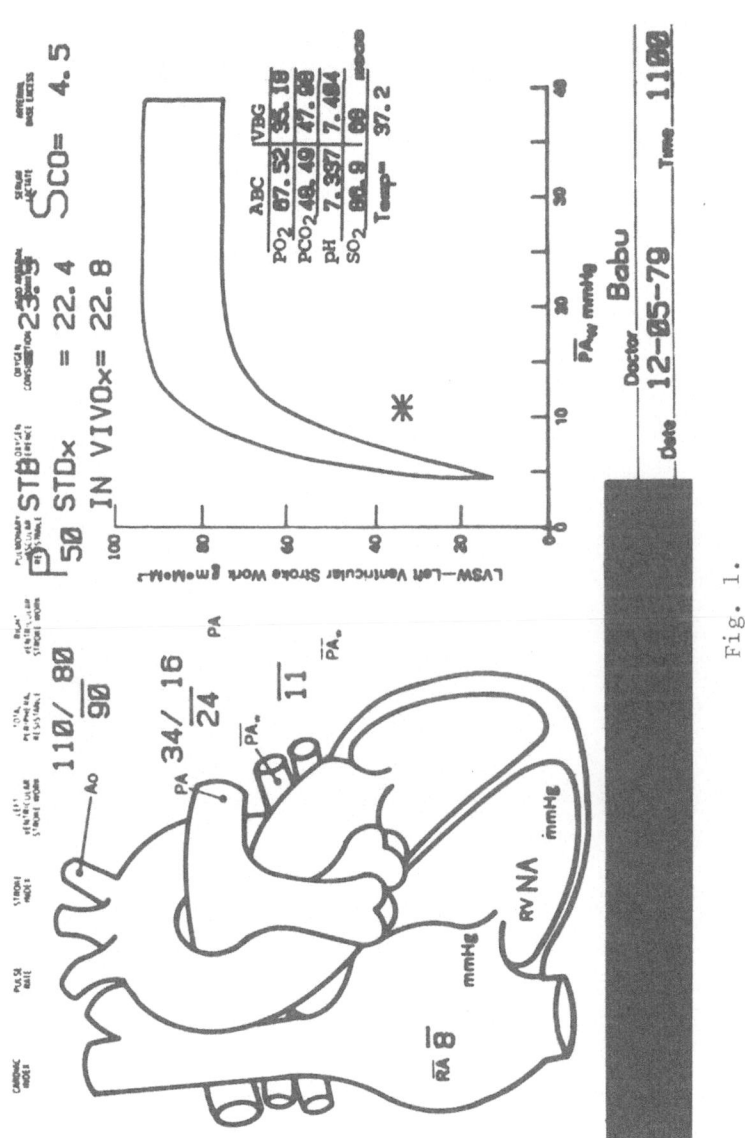

Fig. 1.

REFERENCES

1. J. D. Cohn and L. R. M. Del Guercio, Clinical Applications of
    Indicator Dilution Curves as Gamma Functions, <u>J. Lab. Clin.
    Med</u>. 69:675 (1967).
2. A. L. Gudwin, C. R. Goldstein, J. D. Cohn and L. R. M. Del
    Guercio, Estimation of Ventricular Mixing Volume for Pre-
    diction of Operative Mortality in the Elderly. <u>Annals of
    Surgery</u>, 168:183 (1968).

DATA SYSTEM FOR PATIENTS REQUIRING ARTIFICIAL VENTILATION

P. M. Osswald, H. J. Bender, H. J. Hartung, R. Klose
H. Lutz

The Department of Anaesthesia and Reanimation
Faculty of Clinical Medicine, Mannheim
University of Heidelberg
West-Germany

A computerised data register and processor used in the care of mechanically ventilated patients should fulfil the following functions:

- Optimal monitoring of the ventilation of every patient.

- The connection of the parameters of the mechanics, hemodynamics and bloodgases to the characteristics of each patient.

- Alphanumeric and graphic presentation of the registered and calculated data.

A programme system was developed to record at a variable time interval 30 items of data, "off line". From this data there were 10 derived values.

In the data output a particular value is laid on graphic presentation. Values can be represented singly or as means, so that one can find the average for one patient, for groups and over a period of time.

For revue purposes a survey is possible of several indices which stretch over intervals of 12, 24 and 36 hours or more. A permanent record can be produced.

The task of a computer data register and processor in the management of patients suffering from ARDS should characterise the disease and make an exact analysis possible of the course of the illness and of any change in the lung function.[1]

"Off-line" measurements must always be used whose "output" is successful first of all only with the help of invasive measuring techniques and which are therefore less suitable for a continuous recording lasting for days.

To compare several patients a view of up to 3 x/y plots on the screen at the same time is available. The mean values of groups of patients of a similar type (e.g. shock lung, pneumonia) are therefore possible and so characteristics of the course of illness (for example lung mechanics) can be demonstrated.[2]

REFERENCES

1. Y. Shimada, I. Yoshiya, U. Tanaka, S. Sone, M. Sanurai, Evaluation of the progress and prognosis of ARDS, Chest 75 (1979) 2.
2. A. Steven, B. G. Ronald, Computers assistance in assessment and management of mechanical ventilation, Norwalk. Conn.(1979).

NEW ASPECTS IN MONITORING OF PATIENTS WITH MULTIPLE TRAUMA IN A

COMPUTER ASSISTED ICU

H. Junger, K. van Deyk, R. Weinmann, E. Epple, M. Kopp,
R. Schorer

Institute of Anaesthesiology, University of Tübingen
Calwer Str. 7, D-7400 Tübingen
West-Germany

Patients who have suffered multiple trauma are usually endan-
gered by cardiopulmonary complications. It is important to assess
all data as early as possible and analyse trends by use of a
differentiated monitoring system. Our computer assisted intensive
care unit offers these facilities.

We have explored:

1.   The possibility of computer assisted synoptic display
     assessment of pulmonary and hemodynamic indices in patients
     with trauma in the early and late post traumatic period.

2.   In patients undergoing open heart surgery we examined the
     relationship between the measurements available and the
     degree of variation in heart rate.

For monitoring we used the Hewlett Packard 5600 A Intensive
Care System which allows sampling of the vital data with a rate of
two values per minute. 8-10 times per day cardiac output (Swan-
Ganz-thermodilution-catheter) and pulmonary capillary wedge pressure
were measured, and arterial and venous blood gases were analysed.
From the pulmonary, haemodynamic and blood gas data several different
haemodynamic and pulmonary derived indices were calculated. 18
patients were studied.

In all cases we observed a deterioration of the haemodynamic
indices showing a reduction of the cardiovascular function. There
were always corresponding changes in the heart rate. A reduction
of variability of heart rate indicates a serious alteration of the

cardiovascular system. In those patients with adequate compensation of severe pulmonary injury (ARDS), an improvement of the
cardiovascular performance correlates well with an increase of
heart rate variability.

Our findings suggest that it may be possible to estimate the
cardiac performance by study of heart rate variability. By use of
a computer generated patient chart determination of this relationship is possible, but only in a qualitative manner. We hope,
however, to be able to quantify this in future. Reduction of heart
rate variability is defined as spontaneous variation in heart rate
of less than $\pm$ 5%. It is important to use samples with a rate of
not less than two values per minute and to display and to plot the
non-reduced data.

MONITORING OF LUNG FUNCTION AND HEMODYNAMICS AFTER CARDIAC SURGERY

BY USE OF A COMPUTER ASSISTED ICU

K. van Deyk, H. Junger, F. Münch, M. Kopp, E. Epple,
R. Schorer

Institute of Anaesthesiology
University of Tübingen, Calwer Str. 7
D-7400 Tübingen, West-Germany

Because of various influences of extracorporeal circulation patients undergoing open heart surgery have to be specially monitored. Complications of lung function, heart and circulatory failure in the post-operative phase require an intensive and continuous monitoring of the vital parameters. The assessment and display of vital parameters for estimation of the actual cardiovascular status and for trend determination is simplified in a computer assisted intensive care unit.

The purpose of this study was to determine patterns of pulmonary and a hemodynamic parameters after cardiac surgery. Based on single observations our interest was directed to the variability of heart rate which seems to allow an estimation of cardiac performance.

Our patients have been monitored with the Hewlett Packard 5600, an intensive care system, sampling the vital parameters at a rate of two values per minute. Additional to continuous assessment of the vital parameters cardiac output (Swan-Ganz-thermodilution-catheter) and pulmonary capillary wedge pressure were measured, and arterial and central venous gases were analysed. Based on these data, several different hemodynamic and pulmonary parameters were calculated. By this procedure we obtained experience in 18 patients.

In all patients we observed characteristic changes in heart rate variability. This became evident in an increase of the spontaneous heart rate variability in the post-operative phase demonstrating an increase in the cardiac performance (according to improved hemodynamic parameters).

175

We could also show such a relationship between heart rate
variability and hemodynamic parameters in patients with prolonged
and complicated post-operative recovery period. A good example is
a patient who suffered from an acute myocardial infarction just
after aortocoronary bypass operation. In the early post-infarct
period a low output syndrome correlated with severe restriction of
heart rate variability. With improvement of cardiac performance
normalisation of the hemodynamic data and of the heart rate variability occurred.

The findings of our study imply that the heart rate variability
seems to be a suitable non-invasive vital parameter for the determination of cardiac performance. According to our experiences we
can define the reduction of heart rate variability as a spontaneous
variation in heart rate of less than 5%. To obtain such information
it is however, necessary to use computer systems, which allow the
display of non-reduced data with a sample rate of not less than two
values per minute. The aim of further studies will be to quantify
the relation of heart rate variability to the cardiac performance.

CHAPTER 7

CLINICAL APPLICATIONS OF COMPUTERS IN THE
INTENSIVE CARE UNIT

DEAD SPACE ANALYSIS DURING DIFFERENT VENTILATOR SETTINGS:

USE OF THE SINGLE BREATH TEST FOR CO_2

R. Fletcher* and B. Jonson**

Departments of Anaesthesia* and Clinical Physiology**
University of Lund, S-221 85 Lund
Sweden

The components of physiological dead space ($V_D$phys) were studied during anaesthesia with IPPV for non-thoracic surgery in 87 supine adults. They received 65% nitrous oxide in oxygen, opiates and neuromuscular blockers. We used a Siemens-Elema CO_2 Analyser 930 with a Servo 900 B ventilator to measure expiratory flow ($\dot{V}_E$) and expired P_{CO_2} ($P_E CO_2$). The two signals were recorded on tape at the same time as a sample of blood was taken for gas analysis.

The lungs were ventilated with "small" and "large" tidal volumes (mean 0.45 and 0.75 l, respectively) at which the mean respiratory frequencies were 17 and 9.5 min^{-1}. The ventilatory pattern was square-wave inspiratory flow with end-inspiratory pause (25% and 10% of cycle time, respectively).

The signals were later replayed to a computer (Digital PDP-8) which integrated $\dot{V}_E$ to give expired volume. Relating the instantaneous values for $P_E CO_2$ to this yields the single breath test for CO_2 SBT-CO_2, which was presented on an X-Y recorder together with a statement of the volume of CO_2 in each breath, obtained by integrating the instantaneous products of $\dot{V}_E$ and $P_E CO_2$.

The tracings were corrected for a small analyser delay, and the effect of nitrous oxide on the CO_2 signal. From SBT-CO_2 we obtained tidal volume, V_T, and its components airway dead space ($V_D$aw), and

Acknowledgement: This study was supported by the Swedish Medical Research Council grant no. 14X-02872 and the Swedish National Association against Chest and Heart Diseases.

179

alveolar tidal volume (V_Talv), and also the slope of phase III (the alveolar plateau), which we relate to the mean P_ECO_2. In conjunction with $PaCO_2$, we used the volume of CO_2 in each breath to calculate the part of V_Talv that did not take part in gas exchange, the alveolar dead space (V_Dalv), and V_Dphys.

At small V_T we found that:

V_Daw (ml) = -168 + 145 x body height (m) (RSD = 19, r = 0.59)

The mean V_Daw was about 80 ml at both settings. V_Daw/V_T was 0.18 at small and 0.10 at large V_T. Median V_Dalv was 71 ml at small V_T and 107 ml at large. Median V_Dalv/V_Talv was 0.20 at small V_T (90% of patients lay between 0.07 and 0.39) and 0.14 at large (0.06 -0.34). Thus median V_Dphys was 168 ml at small V_T (102-238) and 192 ml at large (112 - 323) and V_Dphys/V_T was 0.36 at small V_T (0.23 - 0.51) and 0.25 at large (0.15 - 0.42). Phase III slope was smaller at large V_T (p<0.001).

The lower V_Daw/V_T and, more important, lower V_Dalv/V_Talv during large V_T - low frequency ventilation reflect a) increased V_T itself, which gives a more even global distribution of inspired gas,[1] b) increased time for inspired gas distribution between units[2] with different mechanical properties, c) increased time for diffusive gas mixing within terminal units.[3] These three mechanisms probably reduce V_Dalv/V_Talv by allowing better ventilation of dependent, well perfused lung areas and by improving gas exchange within the terminal respiratory units.

Most of the variation in V_Dphys was a result of V_Dalv. V_Dalv/V_Talv was significantly correlated to indicators of airways obstruction such as the flow continuing at the very end of expiration, and the slope of phase III (p<0.01). The reduction in V_Dalv/V_Talv at large V_T implies that time dependent causes of V_Dalv are important. These are uneven distribution and incomplete diffusive mixing of inspired gas, conditions which are otherwise associated with obstructive airways disease.

The findings imply that, airway pressures permitting, large tidal volumes at low frequencies are preferably to small tidal volumes at high frequencies. Further analysis of $SBT-CO_2$ during anaesthesia/ IPPV is in hand and will be reported shortly.

REFERENCES

1. K. Rehder, A. D. Sessler and J. R. Rodarte, Regional intrapulmonary gas distribution in awake and anaesthetised-paralysed man. J. Appl. Physiol. 42:391 (1977).
2. L. Jansson and B. Jonson, A theoretical study on flow patterns of ventilators, Scand. J. Resp. Dis. 53:237 (1972).

3. G. Cumming, A paradox in the ventilation/perfusion hypothesis, in "Mechanics of airways obstruction in human respiratory disease." Tygerberg ed., Balkema, Cape Town (1978).

A NEW SIMPLE AND QUICK METHOD FOR BEDSIDE ESTIMATION OF FUNCTIONAL

RESIDUAL CAPACITY

W. Petro, W. Dams, J. A. Nakhosteen, V. Korn,
N. Konietzko

Dept. of Internal Medicine and Lung Function Diagnosis
Ruhrlandklinik, D-4300 Essen 16
West-Germany

Bedside functional testing of the lung is almost entirely limited to pneumotachography and alveolar and blood gas analysis. The diagnostic nature of such testing is, however, limited.

A modification of the traditional oscillation method was introduced by Korn,[1] which after development of a practical apparatus, gained widespread use. The respiratory impedance, expressed by resistance (R_{os}) and the phase angle (Q) can be read from a special x-y plot. The small device available (Siregnost FD 5) used a constant stroke oscillation generator and a defined reference impedance formed by a tube with low resistance to breathing (Fig. 1). Thus the only parameter left to be measured is the oscillatory pressure. An accessory to this apparatus also allows measurement of He-density linear from 0 to 100% (Fig. 1). Therefore it is possible to measure important values describing actual lung function, the R_{os}, Q and functional residual capacity (FRC). The physical principle on which this analysis is based is the following: a steady oscillatory current causes pressure changes on a diaphragm. These changes are proportional to the gas density.[2] Measuring procedure is identical to that in the traditional rebreathing method requiring a katharmeter for analysis of helium. There are two differences: gas mixing is facilitated by forced breathing.

To evaluate this new method we compared the results to traditional helium gas mixing and bodyplethysmography in 145 subjects.[3]

The correlation between FRC estimated by the new oscillation technique (FRC_{os}) and FRC estimated by traditional helium gas mixing procedure (FRC_2) is close (r = 0.89) but shows a typical

183

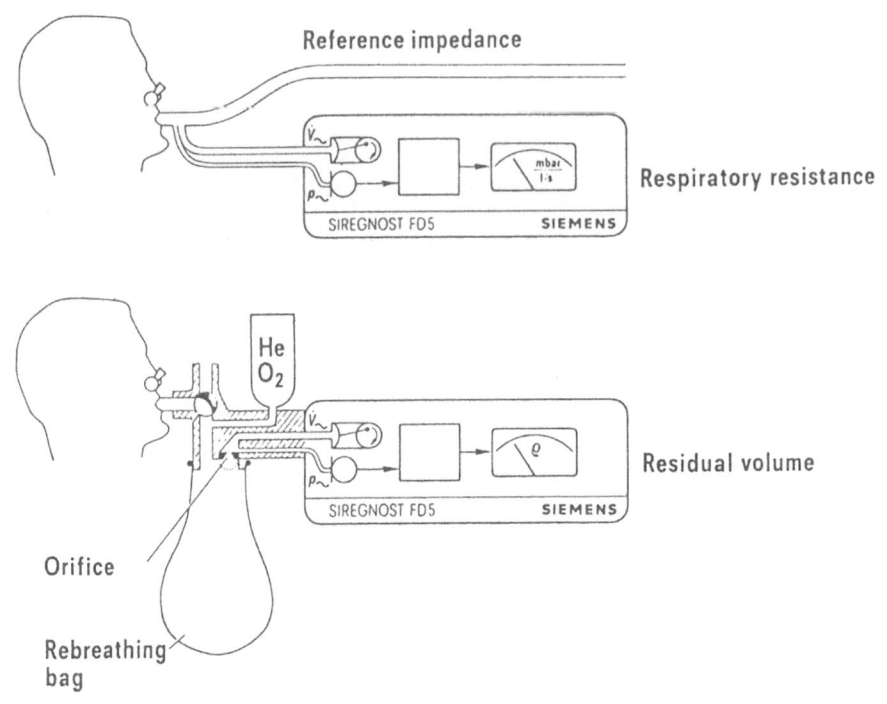

Fig. 1.   Combined apparatus (Siregnost FD 5) for measurement of
respiratory impedance and residual volume.

deviation of the regression line toward the x-axis (y = 0.85x + 0.26).
From that one can conclude, that FRC over records at higher FRC.   In
comparison with bodyplethysmographic FRC (FRC_{box}) = 0.96 + 0.29;
r = 0.88.

The reproducibility of FRC was investigated by paired measurement in 10 normal subjects and was estimated as 97.8%.   Wash-in time
in the traditional helium method is 185 ± 112 s but in the new method
it is only 61 ± 21 s.

The examined side effects of the quick rebreathing manoeuvre
we measured the P_aO_2 before and immediately after gas mixing.
Unexpectedly P_aO_2 increased slightly in obstructive disorders (from
9.66 ± 1.44 to 10.10 ± 1.64 kPa) and decreased in normal subjects
while rebreathing (from 11.89 ± 0.87 to 10.64 ± 1.06 kPa).

The P_aCO_2 is unchanged in patients with airway obstruction and
increased in normals.   The new method is suitable for bedside testing in cases of severe respiratory insufficiency or after operation.

Practicability was tested in the bedside measurement of 12
patients recovering from thoracotomy to assess the effect of

fibreoptic bronchial lavage combined with intrabronchial application of 3 mg of Salbutamol. Gas mixing time in these patients was 70 $\pm$ 23 s   The test was well tolerated in all cases even on the first day after operation. On average the specific conductance increases immediately after lavage and reflects the therapeutic effect (ΔsGaw = 0.143 $\pm$ 0.047 (kPa . L)$^{-1}$). Three hours later a slight decrease occurred (ΔsGaw = 0.045 $\pm$ 0.030 (kPa . L)$^{-1}$).

REFERENCES

1. V. Korn, M Franetzki, K. Prestele, A simplified approach to the measurement of respiratory impedance, Progr. Resp. Res. 11, 1979: 144-161.
2. V. Korn, M Franetzky, U. Smidt, FRC-Bestimmung durch oscillatorische Dichtemessung Wiener Med. Wschr. Suppl. 64 (1980) 4.
3. W. Petro, V. Korn, U. Schmidt, J. A. Nakhosteen, N. Konietzko, Oscillatory He-density measurement - a new procedure for determining intra-pulmonary gas volume, Bull. Pathophysiologie Resp. in press (1980).

VALIDATION OF AN OXYGEN WASH-IN METHOD FOR THE MEASUREMENT OF FUNCTIONAL RESIDUAL CAPACITY IN THE INTENSIVE CARE UNIT

A. M. Benis and T. C. Commerton*
R. R. Mitchell, D. H. McClung and J. J. Osborn**

*Division of Cardiothoracic Surgery
Mount Sinai Medical Center
New York, N.Y.
**Institutes of Medical Sciences
Pacific Medical Center
San Francisco, Ca. U.S.A.

INTRODUCTION

A method for on-line computer monitoring of functional residual capacity (FRC) has been evaluated for use on mechanically ventilated patients in the intensive care unit (ICU). The method uses a change inspired O_2 fraction (FIO_2) of 0.2. or greater. It computes FRC from mean inspired and expired tidal volumes and O_2 fractions measured breath by breath before and after the change in FIO_2 [1].

RESULTS

Comparison of oxygen wash-in method with nitrogen wash-out and helium dilution

Then normal subjects had simultaneous measurements of FRC by both O_2 wash-in method with the use of FIO_2 increase of 0.2 and by nitrogen wash-out. Eight of these subjects had also a sequential measurement of FRC by helium dilution. Correlation coefficients of 0.99 and 0.91 were obtained for the O_2- N_2 and O_2- He FRC results respectively. Paired t-tests revealed no significant differences between FRC measured by either of the two methods.

Validation of oxygen wash-in method with balloon model

The accuracy of the O_2 method waseevaluated in the ICU with the use of a physical model of the lung consisting of a balloon and a section of ventilator tubing dead space with an interposed pneumotachograph. The sensitivity of the measurements of FRC to change in dead space volume was determined, in particular the effect of dead space between the balloon and pneumotachograph (proximal dead space) and between the pneumotachograph and ventilator tubing Y-piece (distal dead space). With distal dead space volumes of 0 and 275 ml and five balloon volumes ranging from 290 to 5270 ml, the maximum and average errors in FRC for 0 dead space volume were 80 and 40 ml; for 275 ml of dead space the errors were 290 and 180 ml. Six repeated measurements with a 1890 ml balloon and 0 and 275 ml dead space volumes yielded FRC values of 1840 ± 80 and 2060 ± 30 ml (mean ± SD) respectively. With 275 ml of proximal dead space the overestimate in FRC (approx 3%) was less than with the distal dead space (approx.10%).

The effect on measured FRC of errors in gas transport delay to the oxygen analyser was determined with the balloon model. Error in FRC was found to be relatively sensitive to error in transport delay in the presence of distal dead space: with a 2.5% overestimate in transport delay the overestimates in FRC were approximately 5, 10 and 15% for zero dead space, 275 ml of proximal dead space and 275 ml of distal dead space, respectively. No significant difference was found between FRC values obtained with a 0.5 increase in FIO_2 (wash-in) and a 0.5 decrease in FIO_2 (wash-in) in the absence of dead space. However, with 275 ml of either proximal or distance dead space values of FRC were approximately 20% greater for wash-out than for wash-in.

CONCLUSION

We conclude that FRC may be monitored precisely and accurately in the ICU by the O_2 wash-in method, even in the presence of proximal dead space, with the use of an increase in FIO_2. With attention to proper calibration of the measurement system, FRC can thus be measured repeatedly and non-invasively in mechanically ventilated patients in the ICU. Studies are presently in progress to determine if measurements of FRC in the immediate postoperative period can aid in the prediction of significant complications of the respiratory subsystem.

REFERENCE

1. D. H. McClung and R. R. Mitchell, Monitoring of functional residual capacity in an intensive care unit, San Diego Biomedical Symposium 15:403 (1976).

2. D. M. Gomez, A mathematical treatment of the distribution of tidal volume throughout the lung, <u>Bull. Nat. Ac. Sc.</u>, 49:312 (1963).
3. W. Weeks, Numerical inversion of Laplace Transforms using Laguerre functions, <u>J. Assoc. Comp. Mach.</u>, 13:419 (1966).

CONTINUOUS DISTRIBUTION OF SPECIFIC TIDAL VOLUME AND FRC FROM O_2

AND N_2 WASH-IN OR WASH-OUT DATA IN ICU

M. Demeester, M. Pellegrini, Y Delcambre

Medical Computing Centre (CIMHUB)
Department of Physiopathology
Free University of Brussels, Belgium

We propose a practical method for obtaining the continuous distribution function of specific tidal volume from data which can easily and automatically be collected from patients under controlled ventilation.

The test procedure consists ot rapidly changing the inspired concentrations of O_2 or N_2 by at least 25% and keeping the new concentration at a constant level (for 10 minutes or more). Pure O_2 breathing is not required. Both the procedure and the equipment needed have been described previously.[1] We use a Siemens Elema Servoventilator 900, hot wire anemometers for gas flow measurements (Spirolog from Draegerwerk, Lübeck, FRG) and a mass spectrometer for O_2, CO_2, N_2 concentration measurements (Perkin Elmer, Pomona, USA). Data are processed in real time by a Digital Equipment PDP 11-34 computer. It calculates breath by breath the mean inspired alveolar O_2 and N_2 pressures (i.e. the partial pressures at the entrance of the alveolar zone) and the mean expired alveolar pressures (i.e. the partial pressures of gas flowing out of the alveolar zone).

The concept of continuous spectrum of distribution of specific tidal volume was first introduced by Gomez.[2] He showed that a N_2 wash-out curve can be described by a Laplace integral equation. Consequently, the mathematical problem of estimating the continuous distribution of specific tidal volume from N_2 wash-out amounts to finding the inverse Laplace transform of the wash-out data.

So far no practical nor reliable general solution has been published. We propose here a method which appears to solve both the theoretical and practical aspects of the problem.

We apply a numerical inversion procedure of the Laplace transform which yields excellent results, i.e. 6 exact decimal digits
when compared to the error-free analytical solution obtained in
specific instances. The procedure is derived from Weeks;[3] it is
based on Laguerre functions expansion; it computes the inverse
Laplace transform of an analytical expression which is first to be
fitted to the raw wash-out data. The first step of the procedure is
critical in as much as it considerably reduces the consequences of
the ill-conditioned character of the Laplace integral (i.e. small
errors on the data produce large errors on the distribution function). The analytical formula was recently proposed to us by Gomez;
it can describe normal or pathological wash-out curves of any type.
We use a "complex box" method to estimate the three paramaters of the
formula.

The proposed method for estimating the continuous distribution
of ventilation assumes that the test procedure does not modify the
alveolo-capillary N_2 and O_2 exchange; but this is not the case. We
correct for N_2 flow from blood and tissue and for additional O_2
uptake by using a mathematical model of N_2 body stores and O_2 exchange. The method also assumes that the wash-out - or wash-in
curves result from a perfect step change in inspired alveolar gas
concentration. In order to comply with this prerequisite we compute
the perfect step response by correlating the observed wash-out data
with the actual inspired alveolar pressure.

FRC can easily be obtained from the same data. The complete
procedure has been checked in the following manner.

- to recover the wash-out curve directly from the estimated
  distribution function
- to recover the theoretical distribution function of ventilation imposed on a mathematical model of O_2- CO_2 and N_2 exchange and transport throughout the body; the model generates artificial wash-out data, which are then processed as
  if they were real
- to observe the reproducibility of the distribution functions
  calculated from series of O_2 and N_2 wash-in and wash-out
  tests performed in subjects under controlled ventilation.

REFERENCE

1. M. Demeester, Ph. Grevisse, Ch. van der Velde, K. Jank, Y. Del-
    chambre, S. Ozkan, A. Swietochowski, P. Mundeleer, Realtime
    and interactive control of O_2- CO_2 transport in mechanically
    ventilated patients in ICU, Computers in Cardiology, IEEE,
    September 1978, 89-96.

ARTERIAL CO_2 TENSION ESTIMATED NON-INVASIVELY FROM END FORCED EXPIRED CO_2 TENSION

M. Berthon-Jones, T. G. Nash, R. Simmul and
R. A. Vandenberg

Respiratory Investigation Unit
Royal North Shore Hospital
New South Wales, Australia

A rapid non-invasive technique has been developed for estimation in the clinical setting of arterial CO_2 tension ($PaCO_2$) from end forced expired CO_2 tension ($PefeCO_2$), and adapted for use in paralysed and ventilated patients in severe respiratory failure.

In the present study, three important modifications have been made to a previously described technique,[1,2] in which the subject expires forcibly to residual volume as a continuation of a normal expiration. Firstly, an operator controlled inspiration blocking valve is added to the mouthpiece valvebox, and is manually closed at the start of the forced expiration. This prevents an abnormally large inspiration, common in confused or less cooperative subjects, prior to or during the expiration. In four normal subjects $PefeCO_2$ was on average 12 mmHg lower after a large inspiration than that it was after a normal inspiration. The inspiration blocking valve prevents this from occurring. Secondly, in sedated or paralysed subjects incapable of performing the forced expiration to residual volume, manual chest compression is performed by the operator for ten seconds immediately after closing the blocking valve. The compression is similar to that by the physiotherapist for mobilization of sputum. Thirdly, an empirical correction for the effects of low ventilation: perfusion regions has been developed.

The accuracy of the technique was assessed by simultaneous arterial puncture under local anaesthesia in 127 subjects. Of these 51 had severe respiratory failure, 43 of whom required constant respirator support at the time. The remaining 69 had moderate or mild respiratory disease. Seven subjects had had major

193

pulmonary embolus within the previous 48 hours. This condition and severe shock are known to affect the $PaCO_2$ - $PefeCO_2$ difference.[1] In the seven with recent major embolus, mean $PaCO_2$ - $PefeCO_2$ difference was 10.7 mmHg (2-22 mmHg), whilst the six subjects with older emboli showed differences in less than 6 mmHg. Thus recent major embolus or severe shock are likely to cause significant error. In the remaining 120 subjects, the largest raw $PaCO_2$ - $PefeCO_2$ difference was 11.8 mmHg (mean 3.3 mmHg). In the 69 patients without severe respiratory failure, the largest difference was 6.8 mmHg only. In the 51 with severe failure, physiologic deadspace, alveoloarterial oxygen tension difference (P_AO_2- PaO_2), and in 18 subjects, physiologic shunt, were measured. Deadspace showed mild positive correlation (R = 0.5) with $PaCO_2$ - $PefeCO_2$ difference. AaDo2 showed R = 0.51. Shunt showed R = 0.82, suggesting low ventilation: perfusion ratio regions accounted for much of the systematic error. Multivariate linear regression of $PaCO_2$ against $PefeCO_2$ and shunt halved the standard error of predicting $PaCO_2$ to 1.6 mmHg. AaDO2 (mean 350 mmHg) was estimated from ear oximetry, with assumed normal pH and respiratory quotient. Multivariate linear regression yielded $PaCO_2$ = 1.0 $PefeCO_2$ + 0.010 $AaDO_2$ - 0.05 mmHg, with R = 0.94, largest error 8 mmHg, 3.3 mmHg standard error.

Particularly when supplemented with ear oximetry, $PefeCO_2$ using the inspiration blocking valve and chest compression if required, provides a clinically adequate estimate of $PaCO_2$ and replacement for repetitive arterial sampling in a wide patient group.

REFERENCES

1. L. Hatle & R. Rosketh, The Arterial to End Expiratory Carbon Dioxide Tension Gradient in Acute Pulmonary Embolism and Other Cardio-Pulmonary Diseases, Chest 66:352 (1974).
2. P. Toulou & P. Walsh, Measurement of Alveolar Carbon Dioxide Tension at Maximal Expiration as an Estimate of Arterial Carbon Dioxide Tension in Patients with Airway Obstruction, Am. Rev. Resp. Dis. 102:921 (1970).

BREATH-BY-BREATH END-TIDAL CARBON DIOXIDE ANALYSIS FOR PATIENTS ON INTERMITTENT MANDATORY VENTILATION

J. E. Brimm, R. K. Brienzo, M. A. Knight and R. M. Peters

University of California Medical Center, San Diego
San Diego, California, U.S.A.*

Monitoring of the peak expired (end-tidal) carbon dioxide partial pressure ($PetCO_2$) has been advocated as a method for non-invasively following arterial carbon dioxide partial pressure ($PaCO_2$) specifically and respiratory status more generally.[1,2] In healthy individuals, agreement between $PetCO_2$ and $PaCO_2$ is good; however, large systematic studies of the agreement in patients in intensive care units have not been performed. In spite of the absence of these studies, manufacturers of both mass spectrometers and infra-red analysers for measuring $PetCO_2$ are marketing their systems for long-term, non-invasive assessment of the $PaCO_2$.

The use of $PetCO_2$ for estimating $PaCO_2$ assumes that $PetCO_2$ measures mean alveolar CO_2 which in turn reflects the $PaCO_2$. This model implicitly assumes that the alveolar gas composition is homogeneous and that the pattern of emptying of lung units is uniform. These assumptions are known to be invalid when ventilation and perfusion are mismatched, for example, in conditions such as pulmonary embolism. In order to compensate for whatever ventilation-perfusion mismatch may exist, $PetCO_2$ monitoring systems use the alveolar-arterial CO_2 difference ($P_ACO_2 - PaCO_2$) determined by simultaneous measurements of $PetCO_2$ and $PaCO_2$. This index is assumed to change much more slowly than either $PetCO_2$ or $PaCO_2$. Subsequently, $PaCO_2$ is estimated as the sum of the alveolar-arterial CO_2 difference and $PetCO_2$. The rate of change of ($P_ACO_2 - PaCO_2$) and the use of $PetCO_2$ to indicate the need for repeated measurements of $PaCO_2$ have not been fully defined.

*Research supported by USPHS Grant Nos. GM-25955, GM-17284, and
HL-13172

195

With the use of intermittent mandatory ventilation (IMV), the assumptions underlying the use of $PetCO_2$ to predict $PaCO_2$ are more tenuous. With IMV, patients are permitted to breathe spontaneously and are supplemented by large volume ventilator breaths given at preset rates. The volumes and rates of ventilator and spontaneous breaths may differ markedly, consequently the uniform pattern of ventilation assumed when $PetCO_2$ is used to predict $PaCO_2$ does not exist.

In order to assess the effect of IMV on $PetCO_2$ and to substantiate its use as a clinically reliable monitor, we studied a group of patients receiving mechanical ventilation to:
1.   determine the variability between $PetCO_2$ of spontaneous ($PetCO_{2s}$) and ventilator ($PetCO_{2v}$) breaths;
2.   assess the differences between $PetCO_{2s}$ and $PetCO_{2v}$ and $PaCO_2$; and
3.   evaluate the stability of ($P_ACO_2 - PaCO_2$) in patients studied serially.

METHODS

We studied 17 patients receiving mechanical ventilation. Of these, 7 had undergone cardiac surgery, 4 had chronic obstructive lung disease, and 5 had other conditions. Selection of patients was based on the availability of technical personnel for carrying out the study, and not according to their respiratory status. All patients were studied in intensive care units of the University of California, San Diego Medical Center, with the approval of the Committee for Human Investigation of the University of California, San Diego.

Respiratory airflow was measured using a Fleisch type pneumotachograph (Hewlett Packard), airway CO_2 concentrations using a mass spectrometer (Perkin Elmer MGA 1100) and airway pressure with a differential manometer. Signals were sampled for two minutes at 25 Hz. by our respiratory testing programmes which run in a Hewlett Packard 5600A Patient Data Management System.[3] We modified a previously described computer programme which automatically discriminates between ventilator and spontaneous breaths based on the flow and pressure and which calculates spontaneous and ventilator rates, tidal volumes, and minute ventilation, compliance, resistance, and work of breathing. $PetCO_2$ values for each breath were measured and classified as to whether they were from ventilator or spontaneous breaths. The average, minimum, and maximum $PetCO_{2s}$ and $PetCO_{2v}$ values were derived along with $PetCO_2$ averaged over all breaths. After studying the first five patients, we added the capability to plot each breath's $PetCO_{2v}$ against that breath's inspired and expired tidal volumes and the time lag since the previous breath's $PetCO_2$.

Immediately following the measurement of $PetCO_2$, we drew an arterial sample for measurement of PaO_2, $PaCO_2$, and pH. The $PaCO_2$ value was used for comparison with end-tidal CO_2 values and for calculation of (P_ACO_2-$PaCO_2$).

Patients were studied serially whenever possible. Repeat studies were done when the ventilator settings were changed, or once per day if the settings were not changed. Studies were terminated either when the trachea was extubated or when the ventilator settings were not changed for two days.

Data were analysed, using the Statistical Package for the Social Sciences, by paired t-test and regression analysis.

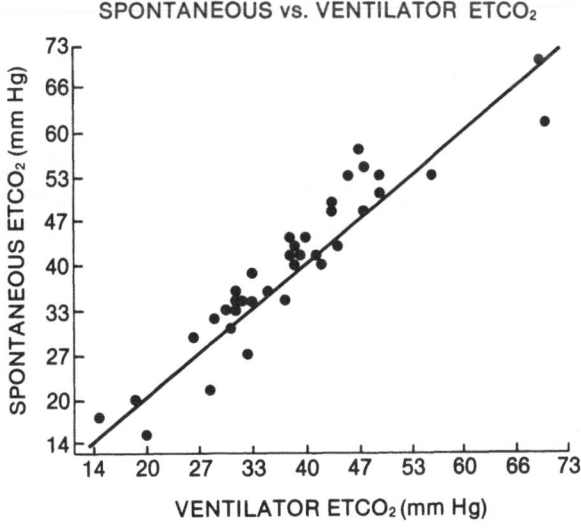

Fig. 1. Spontaneous versus ventilator $PetCO_2$. The solid line is at 45° and represents equality of ventilator and spontaneous values.

RESULTS

There was a total of 63 studies on 17 patients, the maximum number on one patient was eight and four patients were studied only once. The comparison of $PetCO_{2s}$ and $PetCO_{2v}$ is shown in Fig. 1. While $PetCO_{2v}$ and $PetCO_{2s}$ values are linearly related they differ at the 5% level of probability. $PetCO_{2s}$ was almost always greater than $PetCO_{2v}$. In some of the early studies, low $PetCO_{2s}$ values were obtained when patients had very shallow spontaneous expirations that were only slightly larger than 100 ml, the volume we use for classifying the measured tidal volume as a "breath." In these shallow breaths, only dead space gas $PetCO_2$ was measured. When the programme was modified to exclude these "dead space" breaths, $PetCO_{2s}$ was uniformly higher than $PetCO_{2v}$.

Paired t-tests and regressions were done to compare average and maximum $PetCO_{2s}$, $PetCO_{2v}$ and $PaCO_2$. These results pooled from all patients were:

$PetCO_{2s}$	$PetCO_{2v}$	$PaCO_2$	$PetCO_2$	$PetCO_{2v}$ (max)
40.0	38.5	49.0	41.5	45.0

Correlation coefficients are:

$PaCO_2$	-	$PetCO_{2s}$	.815
$PaCO_2$	-	$PetCO_{2v}$	.868
$PaCO_2$	-	$PetCO_{2s}$(max)	.863
$PaCO_2$	-	$PetCO_{2v}$(max)	.881
$PetCO_{2s}$	-	$PetCO_{2v}$	.945

The differences were all significant at the $p = 0.001$ level except for $PetCO_{2s} - PetCO_{2v}$ ($p = 0.005$). The correlation coefficients show that while the difference between $PaCO_2$ and $PetCO_2$ (max) is highly significant, $PetCO_{2v}$ (max) is the most stable predictor of $PaCO_2$.

Serial ($P_ACO_2 - PaCO_2$) values were compared in patients who were studied three times or more. The correlation coefficient was 0.69 but successive values sometimes differed markedly. The maximum change in ($P_ACO_2 - PaCO_2$) was from 0.0 to 13.8. These findings suggest that the assumption of a stable difference is questionable, at least when this index is calculated using $PetCO_2$ as an average of all breaths.

DISCUSSION

When $PetCO_2$ is calculated as an average of all breaths in a time interval, it may differ significantly from the $PaCO_2$. When IMV is used this difference is a result of variation between $PetCO_{2v}$ and $PetCO_{2s}$.

The larger inspired volumes given by the ventilator result in a greater dilution of alveolar P_ACO_2 and a lowered $PetCO_2$. The discrepancy between $PetCO_{2v}$ and $PetCO_{2s}$ can profoundly affect $PetCO_2$ calculated as the average of all breaths depending on the relative rates of spontaneous and ventilator breaths. Often spontaneous rates are much greater than ventilator rates. Thus, $PetCO_{2s}$ is weighted heavily in the breath-averaged $PetCO_2$, yet these spontaneous breaths may contribute very little to overall gas exchange. In several studies we observed that $PetCO_{2s} > PaCO_2 > PetCO_{2v}$ and that breath-averaged $PetCO_2 > PaCO_2$. The most obvious explanation for this finding is that the spontaneous breaths have little effect on the $PaCO_2$. Frequently spontaneous minute ventilation accounted for less than 10% of total minute ventilation and much of this was dead space ventilation.

A further complication of using the breath weighted average is that the spontaneous rate can vary considerably from one two-minute sampling period to another. Although total minute ventilation and $PaCO_2$ were essentially constant, breath-weighted $PetCO_2$ will change. This problem partially accounts for the variability in $(P_ACO_2 - PaCO_2)$ which we observed.

These observations indicate that a better non-invasive measure of $PaCO_2$ during IMV is one which accounts for the volume as well as the $PetCO_2$ level of each breath. By using the average ventilator and spontaneous tidal volumes and rates and $PetCO_{2s}$ and $PetCO_{2v}$, we constructed retrospectively a ventilation-weighted estimate of $PetCO_2$ as

$$PetCO_2 = \frac{RATE_s * TV_s * PetCO_{2s} + RATE_v * TV_v * PetCO_{2v}}{RATE_s * TV_s + RATE_v * TV_v}$$

This ventilation-weighted estimate of $PetCO_2$ showed better correlation with $PaCO_2$ and also reduced the serial changes in $(P_ACO_2 - PaCO_2)$ in each patient. We are now undertaking a second study to verify that ventilation-weighted $PetCO_2$ value gives better agreement with $PaCO_2$.

The implication of this study is that gas analysers cannot be used for monitoring $PetCO_2$ without regard for the pattern of ventilation. For patients receiving IMV, simultaneous flow monitoring seems essential.

REFERENCES

1. J. B. Riker and B. Haberman, Expired gas monitoring by mass spectrometry in respiratory intensive care unit. _Critical Care Medicine_ 4:223-229 (1976).
2. T. C. McAslan, Automated respiratory gas monitoring of critically injured patients, _Critical Care Medicine_ 4:255-260 (1976).

3.   C. M. Janson, J. E. Brimm and R. M. Peters, Instrumentation and
     automation of an ICU respiratory testing system, in: "Computers
     in Clinical Care and Pulmonary Medicine," S. Nair, ed., Plenum
     Publishing, New York (1980)

AN INTERACTIVE COMPUTER PROGRAMME FOR WEANING FROM MECHANICAL VENTILATION

S. Meij, V. v.d. Borden, O. Prakash

Thorax Centre, Erasmus University Hospital
Rotterdam
The Netherlands

The aim of the present study was to test a computer algorithm to obtain a smooth and graded manner to wean patients from the ventilator.

A Servo Ventilator 900B incorporates transducers for the measurement of airway pressure and inspiratory and expiratory flow. Since the ventilator is electronically regulated, digital signals indicating inspiration, expiration, trigger, sigh, or spontaneous ventilation are also available. A carbon dioxide analyser (SE 930) is connected to the ventilator.

The computer system used is a bedside microcomputer monitoring system, developed in the Thorax Centre: the Unibed system. It is capable of monitoring E.C.G., hemodynamic pressures, temperature, fluid output, cardiac output by thermodilution, and respiration.[2,3] It consists of four physically identifiable sub-units: the unibox, the video display unit, the command module, and the coded plug. The unibox itself houses the LSI-11 microcomputer, monitoring place for the display, and the command modules and pockets for the coded plug. The video display is a standard 625-line video system and is used to show up to four signal traces or 32 lines of text or any suitable combination of the two. The command module contains, among other things, four general-purpose signal-conditioning ampli-fiers, an A-D converter, and a 32-character programmable touch-sensitive LED display ("touch-line"). The plugs are special-purpose, in that each one contains coding indicating the function of the attached transducers. Thus, the plug with its sockets and the command module replace the traditional sub-unit by a general-purpose module with a specialised plug which identifies the function to be performed and the programmes to be run by the

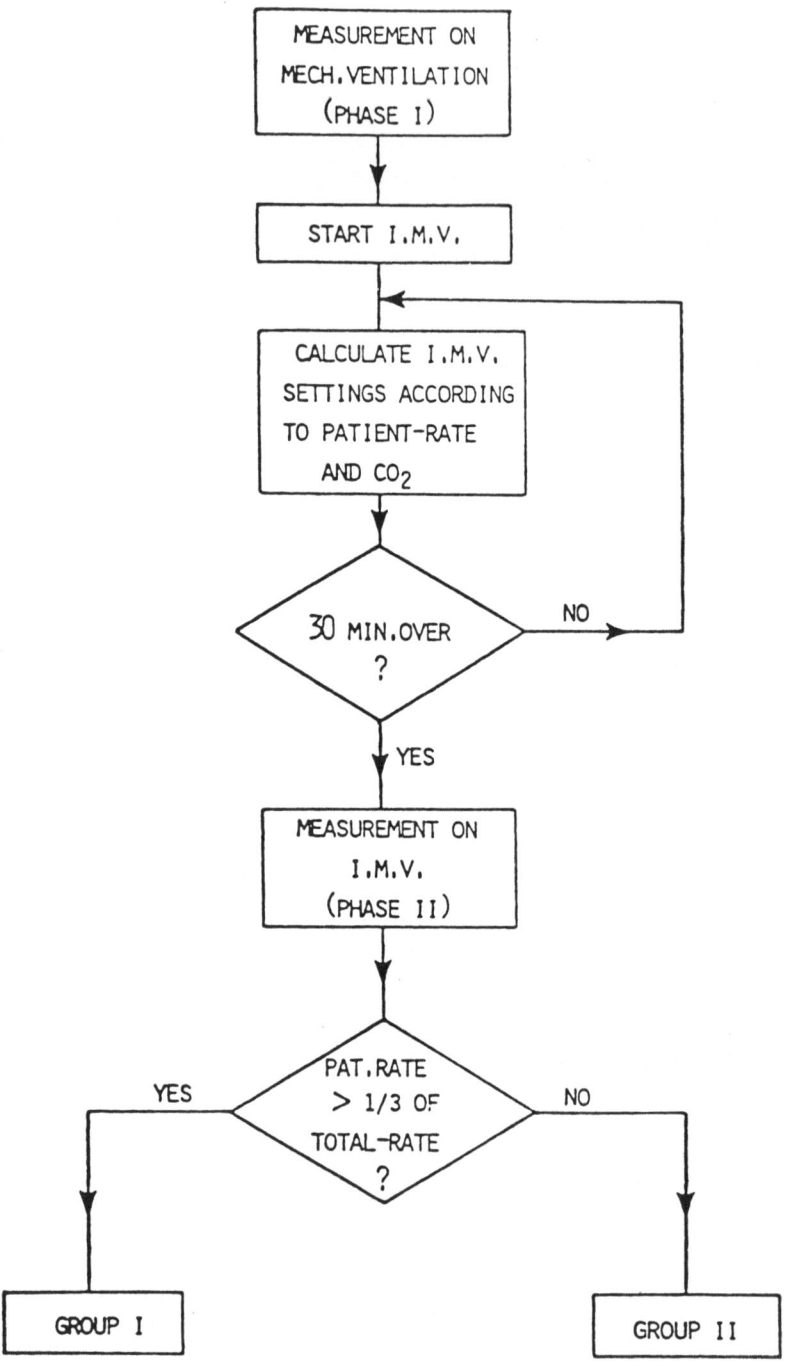

Fig. 1. Flow diagram of protocol for IMV study.

microcomputer. The touch-line in turn replaces the switches, knobs, and dials of the usual front-end by direct communication between the two.

A detailed description of the connection of the Servo ventilator with the unibox follows. The three analog signals (airflow, airway pressure, and airway carbon dioxide) are stored with 12 bits accuracy. The remaining four most significant bits can be used for digital information. These four bits of the airflow are used for the respiration plug identification code. The four digital signals delivered by the ventilator (inspiration timing, expiration timing, patient triggering, and sigh function signal) are copied into the four most significant bits of the airway pressure. This means that with each sample the actual state of the breath is defined. Each module can claim one trace on the video display. One can choose a tracing of pressure, flow, or carbon dioxide, or a summary of all the respiratory items computed.

The IMV section of the programme calculates the best setting for the Servo ventilator when the patient is on IMV. Input for this section is tidal volume, total respiratory rate, and trigger rate. The ventilator has three IMV modes: F/2, F/5, and F/10, or can be set on FO, spontaneous ventilation. F/X means that the present ventilatory frequency and minute volume is divided by divisor X. The programme calculates the best divisor X, the ventilator frequency, minute volume, and the optimal inspiration time. A trigger during the IMV period is taken over by the ventilator. If there is no trigger at all during the IMV period, the ventilator generates a breath.

A protocol was set up to test this programme on twenty adult patients who underwent coronary artery bypass surgery (Fig. 1). The status of the patients was relatively good, and difficulty in weaning was not expected. After arrival in the intensive care unit and when nitrous oxide wash-out was completed, measurements were made under mechanical ventilation (Phase I).

Then the weaning procedure started for a period of 30 minutes and measurements were repeated (Phase II). When the patient did not trigger at the beginning of the weaning period, the ventilator was set at F/2 and volume and rate were doubled. When during Phase II the trigger rate was less than 1/3 of the total rate, or when end-tidal carbon dioxide rose above 6 kPa, the weaning procedure was defined as a failure.

Fifteen of the patients (Group I) were successfully weaned, but weaning was unsuccessful with the remaining five patients (Group II), who could not be identified by measurements during Phase I (Table 1).

Table 1. Hemodynamic and gas exchange data

Phase*	Group I - Success (n=15) I	Group I - Success (n=15) II	Group II - Failure (n=5) I	Group II - Failure (n=5) II
LA (mmHg)	10 ± 1	9 ± 2	11 ± 2	10 ± 2
PA (mmHg)	18 ± 2	18 ± 2	21 ± 2	21 ± 3
Art (mmHg)	112 ± 6	102 ± 4	104 ± 6	99 ± 15
CI ($l/min/m^2$)	2.7 ± 0.2	2.8 ± 0.2	2.2 ± 0.4	2.7 ± 0.4
HR (/min)	102 ± 6	102 ± 6	103 ± 6	106 ± 7
MV (l/min)	9.5 ± 0.5 ¶	7.2 ± 0.4	8.0 ± 1.0	8.2 ± 0.9
CO_2 (kPa)	4.9 ± 0.2 †	5.6 ± 0.2	4.7 ± 0.2 §	5.6 ± 0.5
VO_2 ($ml/min/m^2$)	150 ± 17	120 ± 8	134 ± 21	130 ± 5
RQ	0.84 ± 0.01	0.80 ± 0.03	0.79 ± 0.02	0.84 ± 0.02
VD/VT (%)	34 ± 1	37 ± 1	35 ± 2	34 ± 2

Values are mean ± SEM. ¶ $p < 0.001$ † $p < 0.01$ § $p < 0.05$
* Phase I - measurement on MV when N_2O wash-out less than 0.5%
  Phase II - after 30 min on IMV.

IMV was introduced in an attempt to solve the problem of difficult weaning from the ventilator. In T-piece weaning, the mechanical ventilatory support is withdrawn in an all-or-none manner, whereas in IMV weaning, it is withdrawn in a smooth and graded manner. The patients who were marked as failures could not be identified by measurements during mechanical ventilation (Table I). The difference between the two groups was that patients of Group I gradually increased their trigger rate and those of Group II now and then did not trigger at all, causing hypoventilation, which in turn leads to an increase of the IMV.

We believe that "by closing the loop", which in turn requires certain modifications of the ventilator, we can reach an optimal state for gradual weaning. Especially for Group II patients, we feel that if the IMV-period were under computer control, we could have avoided hypoventilation, which was the main cause of failure.

REFERENCES

1. S. Ingelstedt, B. Jonson, L. Nordström, S. G. Olsson, A Servo-controlled ventilator measuring expired minute volume, airway flow and pressure, Acta Anaesth. Scan. Suppl. 47 (1972).
2. L. S. Deutsch, W. A. H. Engelse, C. Zeelenberg, F. van der Voorde, D. C. Hugenholtz, The Unibed patient monitoring system: A new approach for a new technology, Med. Instrum. Vol. II, No. 5, 274 (1977).
3. C. Zeelenberg, W. A. H. Engelse, L. S. Deutsch, A hierarchial patient monitoring computer network, Proc. Computers in Cardiology, IEEE Computer Society, No. 77 CH 1245-2C, pp 439 (1977).

FEEDBACK CONTROLLED VENTILATION

D. R. Westenskow, W. S. Jordan, K. B. Ohlson

Departments of Anaesthesiology and Bioengineering
University of Utah
Salt Lake City, Utah, USA

Feedback control of mechanical ventilation provides continuous adjustment of ventilation in response to the patient's metabolic rate. Arterial blood carbon dioxide concentration ($PaCO_2$) can be held within a desired range despite wide changes in carbon dioxide production ($\dot{V}CO_2$). The input signal to the feedback controller must be wisely chosen to ensure system safety and reliability. End-tidal PCO_2 is often used because of advanced sensor technology, rapid response time and because it is non-invasive. End-tidal PCO_2 becomes an inaccurate control parameter in the presence of a large ventilation-perfusion (V/Q) mismatch, a rapid respiratory rate, certain pulmonary diseases or rapid changes in circulation.[1] Our approach is to use end-tidal PCO_2 for moment to moment control of ventilation but to modulate this control with whole body oxygen consumption ($\dot{V}O_2$) and carbon dioxide production. These slower changing variables related to body metabolism are used to ensure that ventilation is changed only when metabolism changes and is not changed as a result of sudden changes in circulation or V/Q mismatch. $\dot{V}O_2$ and $\dot{V}CO_2$ are also used to indicate the need for an arterial blood measurement of PCO_2.

The ventilator controller is based on the Intel 8085 microprocessor.[2] Sixteen input parameters are sampled with 12 bit accuracy. The sampled data is displayed on an integral CRT in tabular or trend form. A hard copy is produced on a small thermal printer. A D/A converter produces signals for controlling the ventilator.

The Siemens 900B Servo-Ventilator is used with the 930 CO_2 analyser and 940 Lung Mechanics Calculator. Signals measured by the Siemens system and sampled by the micro-processor include carbon dioxide production, ineffective tidal volume, end-tidal CO_2

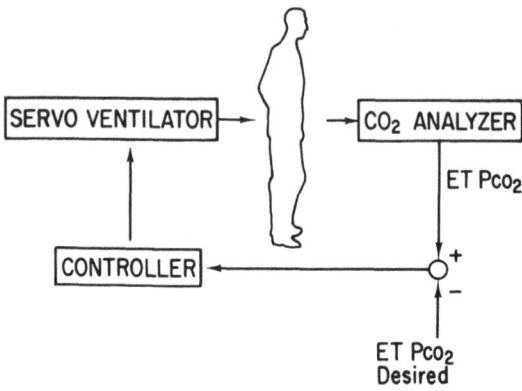

Fig. 1.

concentration, lung compliance, mean airway pressure and peak airway pressure. Oxygen consumption is measured by an oxiconsumeter, a device developed at the University of Utah.[3]

The feedback controller uses proportional, integral and derivative (PID) elements to maintain a constant end-tidal PCO_2. The measured PCO_2 is compared with the desired $PaCO_2$ to produce an error signal to the PID controller. Ventilator tidal volume is adjusted until the error signal is zero. Alveolar ventilation is calculated using a steady state value of $\dot{V}CO_2$ and the equation of Mitamura et al. (IEEE Trans. 18:330). Calculated alveolar ventilation is compared with actual alveolar ventilation and used as a measure of V/Q or change in V/Q. A change in $ETCO_2$ not associated with a corresponding change in $\dot{V}O_2$ and $\dot{V}CO_2$ is not used to modify ventilation.

The system has been evaluated in dogs where severe changes in V/Q were produced by inflating the balloon of a Swan-Ganz catheter or blocking one lumen of a dual lumen endobronchial tube. Changes in $\dot{V}CO_2$ were produced by infusion of sodium bicarbonate. With normal V/Q ratio and increases in $\dot{V}CO_2$ of 60%, the controller maintained arterial PO_2 within 4 torr or the desired set point. When V/Q was altered significantly, the arterial end-tidal PCO_2 difference increased to over 24 torr and the set point had to be reset to compensate. The micro-processor automatically makes the adjustment once the arterial blood PCO_2 is known. Future changes in the calculated alveolar ventilation-measured alveolar ventilation difference are used to indicate need for an additional arterial blood measurement.

REFERENCES

1. Y. Mitamura, T. Mikami, H. Sugaware, C. Yoshimoto, An optically controlled respirator, IEEE Trans. Biomed. Eng. 18:330-337 (1971).
2. W. S. Jordan, D. R. Westenskow, Micro-processor control of ventilation using carbon dioxide production, Anaesthesiology 51:S380 (1979).
3. D. R. Westenskow, D. B. Raemer, D. K. Gehmlich, C. C. Johnson, Instrumentation for monitoring oxygen consumption using a replenishment technique, J. Bioeng. 2:219-227 (1978).

DETERMINATION OF OPTIMUM PEEP: COMPUTERISED ANALYSIS OF THE EFFECT

OF ARTIFICIAL VENTILATION WITH VARIOUS LEVELS OF PEEP

H. Mrochen, W. Kuckelt, R. Dauberschmidt, U. Hieronymi
and Manfred Meyer

Research Department of Intensive Medicine
Friedrichshain Hospital
Berlin, West Germany

An accepted measure in present-day treatment of acute respiratory failure is artificial ventilation with positive end-expiratory pressure (PEEP).[1,2] The fact, that there is no unique "best PEEP" makes the determination of PEEP a recurrent problem in every patient.

For that purpose we used the evaluation function I_p ("index pulmonalis") which was developed for the assessment of gas exchange disturbances in critically ill patients.[3] This function is based on measured arterial blood gases and inspired oxygen fraction, and describes the severity of any impairment of the gas exchange function of the critically ill. The formula of I_p is

$$I_p = 2120 - 2 \cdot \frac{P_aO_2}{F_IO_2} - 3 \cdot \frac{AaDO_2}{F_IO_2} - P_aCO_2$$

The values of I_p are an expression of the physiologically relevant information conveyed by the measured data of patients suffering from respiratory insufficiency.

To find optimum PEEP we use a method developed in 1974 by the French mathematician J. L. Lagrange. As a method of interpolation its resultant curve should coincide with the measured values. We believe, that there should exist at least one maximum, that the resulting function should be an even one, and that there should be a simple and reasonable standardisation of the procedure.

Lagrangean Interpolation led us to a polynomial of 6th degree using the free 7th value at an assumed EEP level of -5 cm H_2O for standardisation.

211

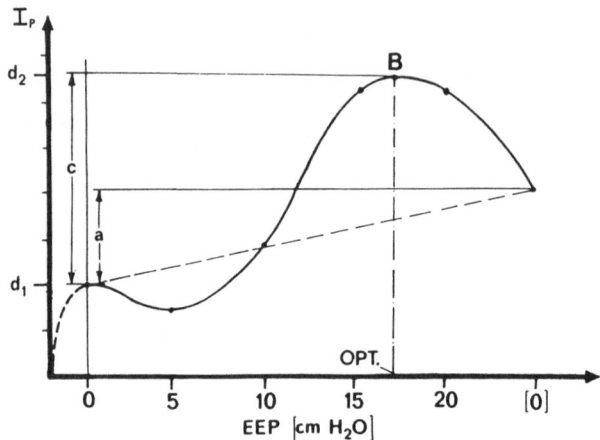

Fig. 1. The graphical display: Gas exchange improvement; a) by selection procedure, c) by OPT peep, B) point of max. advantage, [d_1, d_2] level of gas exchange evaluation function.

In figure 1 the principle and an example for our method are
shown.

The main determinants are: the distance "a" is a representation of the changes in gas exchange capability due to the repeated
measuring procedure. The dashed line between the measuring points
at zero end-expiratory pressure (ZEEP) before and after the measurements gives an impression of the changes happened. The capital "B"
denotes the maximim of the curve, selected as point of the optimum
PEEP level. Distance "c" is the overall improvement by means of
application of optimum PEEP with respect to the basic value. The
scale, in which the changes were presented, is that of the evaluation function I_p. The interval (d_1, d_2) shows for every patient
the region in which his values lie.

Using the generally accepted method for finding optimum PEEP
we derived by means of the evaluation function I_p a graph represent-
ing the physiological response of any patient on the EEP applied.
This graph provides an optimum PEEP level, allows the identification of patients requiring inconvenient high PEEP and points out
patients for whom PEEP is not recommendable. It gives an information on the individual degree of improvement of the gas exchange
function of the lung following PEEP. All calculations, including

those for working out the 104 physiological profiles, one for every individual measurement, were made on a desk-top computer system of Hewlett Packard. For the estimation of the interpolated curve we used the graphical capabilities of the HP 9845, which gives the graph with the marked optimum PEEP within a few seconds. But any desk-top computer with plotter or graphical display should be sufficient.

The principle can be incorporated into monitoring systems for microprocessor controlled ventilation with PEEP. The method presented is a contribution to an objective measure of the effect of PEEP on the gas exchange. It supports the clinical assessment of the physician reducing the risk of misinterpreting clinical data.

REFERENCES

1. J. F. Murray, Mechanism of acute respiratory failure, Amer. Rev. Resp. Dis. 115, 1071-1078 (1977).
2. D. G. Ashbaugh, T. L. Petty, Positive end-expiratory pressure: physiology, indications and contraindications, J. Thorac. Cardiovasc. Surg. 65, 165-170 (1973).
3. H. Mrochen, H. W. Kuckelt, B. Haendly, M. Meyer, An index for assessing the pulmonary function in the critically ill, Crit. Care Med. submitt. f. publication.

EXPERIENCE WITH A COMPUTER - ACTIVATED TREND - DETECTION ALARM FOR MEAN ARTERIAL AND LEFT ATRIAL PRESSURES

T. C. Commerton, A. M. Benis, H. L. Fitzkee, R. A. Jurado
and R. S. Litwak

Division of Cardiothoracic Surgery
Mount Sinai Medical Center
New York, N.Y., U.S.A.

In our cardiac surgical intensive care unit (ICU) cardio-pulmonary monitoring is accomplished with the aid of a central computer (IBM 1800) which routinely samples several parameters at ten minute intervals. For two of the monitored variables, namely mean arterial pressure (MAP) and left atrial pressure (LAP), we have programmed the computer to activate an alarm sequence.

The objectives of the alarm system for use in the ICU were as follows:

1. to establish precise objective criteria under which the nurse must summon the physician to the bedside;
2. to assure that the physician examines a graphical plot of the trends of MAP and LAP before instituting therapy; and
3. to minimize the possibility of missing a slow adverse trend in MAP and LAP.

Untoward clinical events were classified as: Acute Episodes and Adverse Trend Conditions (ATC's). Acute episodes are those events which occur rapidly, on a time scale of 0-15 minutes. ATC represents an event that occurs more insidiously on a time scale of 15 minutes or greater. The alarm system as described here was designed specifically to detect ATC's.

An alarm algorithm was designed to monitor MAP and LAP and to initiate a repeat mode of sampling at one minute intervals if either are outside preset limits. If three consecutive repeated values remain out of limits then the following occurs:

- a bedside light and audible chime sound the alarm;
- the bedside computer keyboard is locked out to other functions until the alarm is reset; and
- a three hour trend plot of MAP and LAP appears on the bedside video monitor.

Based on the results of a pilot study the alarm algorithm was modified to require five consecutive repeated values of MAP or LAP to be out of limits before the alarm was sounded, to allow a thirty minute period of silence following an alarm to permit initiation of therapy.

METHODS

The two alarm algorithms were evaluated in the ICU in separate studies with the bedside light and chime non-operational. The alarm status was recorded every 15 minutes, the alarm categories being: True Positive (TP), False Negative (FN), False Positive (FP), and True Negative (TN), depending on whether an ATC existed and whether the alarm sounded or was appropriately silent.

The patients in the studies were confined to bed. Table 1 illustrates the subgroups of the studies with the three and five minute algorithms.

Table 1. Subgroups of study

	Operative Day		Post-op Day	
No. patients	11	11	12	11
Trachea intubated	11	11	3	3
Hours	61.5	59.25	60.3	65.5

Statistical evaluation of each study was done with the use of the standard terminology of epidemiology.[1] The incidence of alarms and incidence of false alarms were calculated per patient per eight hour shift.

RESULTS AND DISCUSSION

Table 2 summarises the results of the two studies. In the pilot study[2] there occurred 43 TP and no FN's. This is reflected in a sensitivity of 1.00 for both study days. The prevalence of ATC was higher on the operative day (0.13) as compared with the post-operative day (0.04). The incidence of alarms was correspondingly higher on the operative day (5.5) than on the post-operative day. However, the incidence of false alarms was about the same for both study days, that is, approximately 1-2 false alarms per patient per eight hour shift, being well within the tolerable range. The predictive value of the positive test was higher on the operative day (0.75) as opposed to the post-operative day (0.42). This is mainly a result of false alarms of the LAP type which occurred on the post-operative day. In the second study (Table 2) we noted a reduced incidence of alarms as compared with the pilot study. Also evident is a reduced incidence of false alarms on both study days (0.7 vs 1.4 and 1.2 vs 1.8). The reduction of alarms was mainly a result of the elimination of marginal true alarms, when the MAP or LAP was intermittently within limits and then migrated outside of limits. This is also reflected in a reduced effective prevalence of ATC in the second study (0.05 vs 0.13 and 0.03 vs 0.04). In the second study there was a marked reduction in false alarms of the LAP type as a result of patient instability, (the occurrence of an alarm when a patient was turned or suction applied). False alarms from improper levelling of the LA transducer persisted, however. In contrast, the true alarms were most often of the MAP type.

Table 2. Indices for Evaluation

	Operative Day		Post-op Day	
Sensitivity	1.00	1.00*	1.00	1.00*
PV of positive test	0.75	0.71*	0.42	0.44*
Prevalence	0.13	0.05*	0.04	0.03*
Incidence of alarms #	5.5	1.9*	3.0	2.1*
Incidence of false alarms #	1.4	0.7*	1.8	1.2*

Asterisk (*) denotes study with 5 minute algorithm (second study).
calculated per patient per eight hour shift.

Our revised alarm system, employing the five minute algorithm has been in routine operation. We have, on average, two true alarms and less than one false alarm per patient per eight hour shift. In current use, our programme of study calls for the dis-arming of the LAP alarm once respiratory weaning has been initiated, thus eliminating most of the false alarms of the LAP type.

REFERENCES

1. D. J. P. Barker, Practical epidemiology, Churchill Livingston,
    New York, p. 16-21 (1976).
2. A. M. Benis, H. L. Fitzkee, R. A. Jurado and R. S. Litwak,
    A computer activated two-variable trend alarm, Proceedings
    First Annual International Symposium on Computer in Critical
    Care and Pulmonary Medicine, Norwalk, Conn., May 24-26 (1979).

AUTOMATIC ANALYSIS OF CIRCULATORY SHUNTS DURING ARTIFICIAL

VENTILATION

J. R. C. Jansen, J. M. Bogaard, R. Spritzer,
A. F. M. Verbraak, A. Versprille

Pathophysiological laboratory
Department of Pulmonary Diseases
Erasmus University, P.O. Box 1738
3000 DR Rotterdam, The Netherlands

Because of problems of asphyxia, prematurity (respiratory distress) and diseases (meningitis) the lungs of some neonates need to be ventilated artificially. Most of these children are ventilated without knowledge of the cardiac output and circulatory shunts. To evaluate these variables, under different circumstances of artificial ventilation, an animal model (new-born pig) has been developed.

In the case of left-to-right shunt through the foramen ovale a bimodal indicator dilution curve is obtained after bolus injection of an indicator to the venous compartment of the systemic circulation with detection of the concentration time relationship in the arterial part of this circulation. The indicator passing through the foramen ovale will be detected earlier than indicator passing through the pulmonary circulation. To obtain the area of the shunt curve separate from the total dilution curve, several techniques can be used: 1. curve fitting based on empirical formulae, 2. curve fitting based on a semi-logarithmic extrapolation, which uses only a part of the descending limb of a dilution curve, 3. curve fitting based on a local density random walk (LDRW) function, which uses the whole curve and is based on a physical model. The empirical formulae and the semi-logarithmic extrapolation techniques are not based on any real model representation. LDRW function is based on a diffusion or dispersion of indicator particles superimposed on the linear drift of the fluid. Earlier theoretical and experimental investigation have shown that a great majority of curves with various shapes (and noise) can be fitted to this function with great accuracy.[1]

The local density random walk is defined by:

$$c(t) = a(e\lambda/\mu) \, (\lambda/2\pi)^{\frac{1}{2}} \, (u)^{\frac{1}{2}} \, \exp. \, (-\tfrac{1}{2}\lambda(u + u^{-1})$$
$$\text{with } u = \mu(T - T_o)$$

a = area under the primary curve (Fig. 1.)
λ = factor, related to the degree of asymmetry of the curves,
μ = transit time of the median particle,
T_o = zero time of the distribution function related to the
   real time axis.

For the estimation of these indices the Gauss-Newton method is used. The Gauss-Newton subroutine, which is available as a standard routine for non-linear least squares fitting, has been demonstrated to be very insensitive for the starting values of the parameters and for noise. Therefore a simple calculation can be made from the triangle A B C (Fig. 1.) to obtain the starting values of the fit procedure. The truncating point must be indicated on the descending limb of the first part of the shunt curve. At this point it may be assumed that there is no or very little interference from the second part of the curve. From Fig. 1. it is obvious especially in tracing B that good prediction of the area of the primary curve will be

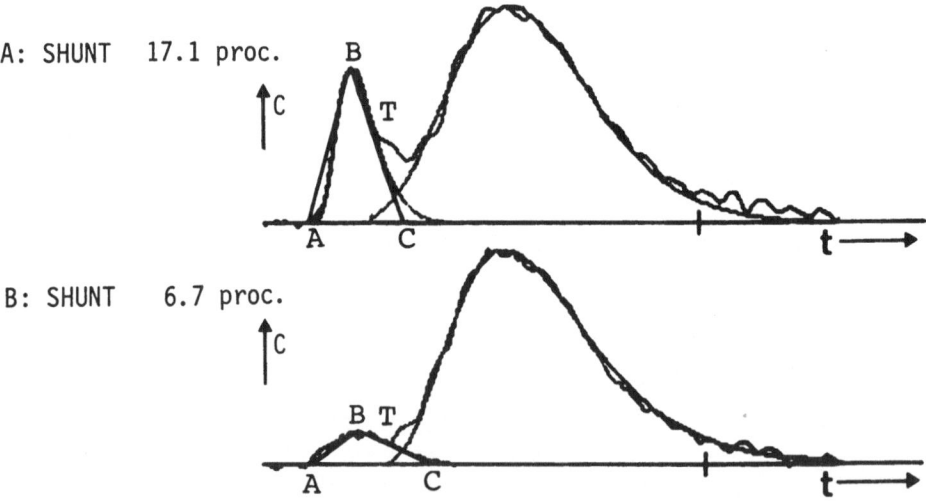

Fig. 1. Two dilution curves obtained experimentally in a patient
         with right-to-left shunt shown together with LDRW fits
         (smooth linear). T is the truncating point. A, B and C
         are the points of the triangle, constructed to obtain
         starting values for the fit procedure.

difficult with semi-logarithmic extrapolation, because not enough information is available in the descending limb of the first primary curve. LDRW-function, however, uses the whole curve as far as the truncating point and makes a good fit possible.

The automatic fit procedure, based on the LDRW model, has proved to be very reliable. The method has great advantages for analyses of shunt curves in clinical practice, both for the speed of analyses and the improving accuracy of the shunt estimates.

REFERENCE

1. J. M. Bogaard, Interpretation of indicator-dilution curves with a random walk model, Diss. Abstr. Int., section C, 1980 (in press; thesis Erasmus University, Rotterdam, 1980 (with a summary in English).

DYNAMIC DESCRIPTION OF THE CARDIAC PERFORMANCE DURING INTERMITTENT

POSITIVE PRESSURE VENTILATION IN DOGS

L. Sibille, F. Fraisse, J. M. Vallois, C. Gaudebout

U. 13 Inserm, Hôpital Claude Bernard
75019 Paris
France

Hemodynamic data change throughout each respiratory cycle (RC) indicating that ventilation and circulation are related. To study this phenomenon, we developed a computerised method that describes coupling between ventilation and circulation in dogs during intermittent positive pressure ventilation. Dogs were anaesthetised and the lungs mechanically ventilated with a tidal volume of 12 ml kg^{-1} and a frequency of 20 per minute. After control measurements, hemodynamic changes were produced by three rates of dobutamine perfusion: 10, 20, 40 $\mu g.kg^{-1}.min^{-1}$. Aortic pressure (AP) was detected by a Cournand catheter (9F) positioned in the ascending aorta; left ventricular pressure (LVP) was detected by transthoracic puncture; airway pressure (AwP) was detected by a Cournand catheter positioned in the trachea. AP, LVP, AwP were stored on magnetic tape after A/D conversion at 200 Hertz.

A computer programme determined, beat to beat, systolic (SAP) and diastolic (DAP) AP, end diastolic LVP (EDLVP), the maximum (Dmax) and the minimum (Dmin) of the isovolumic left ventricular contraction and relaxation phases, the LVP at the Dmax time (PDmax) and Dmin time (PDmin), and the mean AwP throughout each cardiac cycle. Using the computer, we compressed all RC recorded during one minute into a single schematic RC (SRC). In this manner we constructed a continuous representation of any given parameter, although the data from which the curves were formed are discontinuous (one value per cardiac beat) throughout the RC.

For example, SRC of Dmax superimposed on SRC of AwP during infusion and the results for the data calculated during infusion of Dobutamine (10 $\mu g.kg^{-1}.min^{-1}$) are given in figure 1 and the accompanying table. We define respiratory difference (RD) as the distance

223

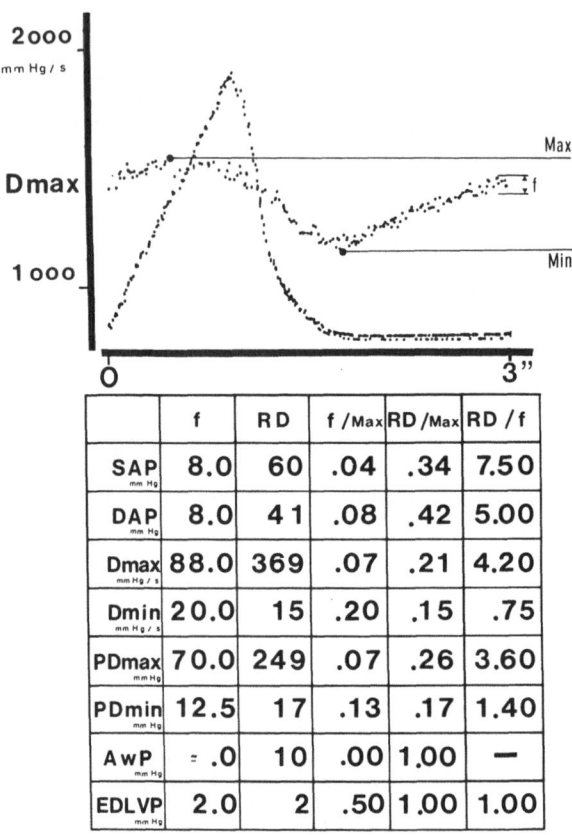

	f	RD	f /Max	RD /Max	RD /f
SAP mm Hg	8.0	60	.04	.34	7.50
DAP mm Hg	8.0	41	.08	.42	5.00
Dmax mm Hg / s	88.0	369	.07	.21	4.20
Dmin mm Hg / s	20.0	15	.20	.15	.75
PDmax mm Hg	70.0	249	.07	.26	3.60
PDmin mm Hg	12.5	17	.13	.17	1.40
AwP mm Hg	=.0	10	.00	1.00	—
EDLVP mm Hg	2.0	2	.50	1.00	1.00

Fig. 1. A schematic respiratory cycle showing D_{max} (mmHg/sec) and
         AwP (mmHg), plotted against time elapsed between the onset
         of systole and the beginning of the corresponding RC, dur-
         ing Dobutamine infusion (10 µg/kg.min); the table shows
         results computed for eight hemodynamic variables during
         the same rate of infusion. For definitions see text.

between the maximum and the minimum values is a SRC and fluctuation (f) as the dispersion of points at a given moment during a SRC.

All the f/max ratios were less than 10% except for PDmax, PDmin and EDLVP. However, the EDLVP fluctuation is only 1 mmHg, or twice the overall precision of the pressure measuring system. The large f values of PDmax and PDmin may be accounted for by errors of computation. Our computer, small in size, imposed simple and short routines for Dmax and Dmin. More sophisticated routines might reduce f of PDmax PDmin. The RD or SAP differs from the RD of DAP; similar variations occurred in comparisons of Dmax to Dmin, and of PDmax to PDmin. These RD variations from one moment to another of a pressure contour indicate that AP and LVP pulses vary in form throughout RC. This variation depended on the rate of drug infusion. Moreover the values of one parameter varied markedly from one cardiac beat to the next; for example, the ratios of the pressure pulse values from one cycle to the following may be as great as 20%.

These extensive variations of AP and LVP pulse contours from one cardiac beat to the following allow us to conclude that circulatory pressure is not periodic and that ventilation and circulation are not linearly interrelated. Our method provides a precise tool with which to analyse such interactions between ventilation and circulation.

AUTOMATED CONTINUOUS RESPIRATORY QUOTIENT TEST

R. M. Peters, R. K. Brienzo and J. E. Brimm

Department of Surgery
University of California School of Medicine
San Diego, California, U.S.A.*

Determination of instantaneous continuous respiratory quotient (RQ) of a single breath has been proposed by Wagner and colleagues[1] as a non-invasive method of assessing ventilation perfusion (V/Q) abnormalities. To perform the calculations of continuous RQ, air flow and expired carbon dioxide, oxygen and nitrogen must be recorded during a single breath. We sampled each signal 25 times per second and then used the following formula to calculate RQ at each sample interval.

$$RQ = \frac{P_E CO_2}{\dfrac{P_{I_{O_2}} \cdot P_{E_{N_2}} - P_{E_{O_2}}}{P_{I_{N_2}}}}$$

where $P_{I_{O_2}}$ = partial pressure inspired oxygen

$P_{E_{O_2}}$ = partial pressure expired oxygen

$P_{I_{N_2}}$ = partial pressure inspired nitrogen

$P_{E_{N_2}}$ = partial pressure expired nitrogen

$P_E CO_2$ = partial pressure expired carbon dioxide

The calculated RQ values are then graphed as a function of expired volume (Fig. 1). The shape of this curve is sensitive to the form, rate, and depth of inspiration ($\dot{V}O_2$) and expiration (VCO_2),

*Research supported by USPHS Grant Nos. GM-17284 and HL 13172.

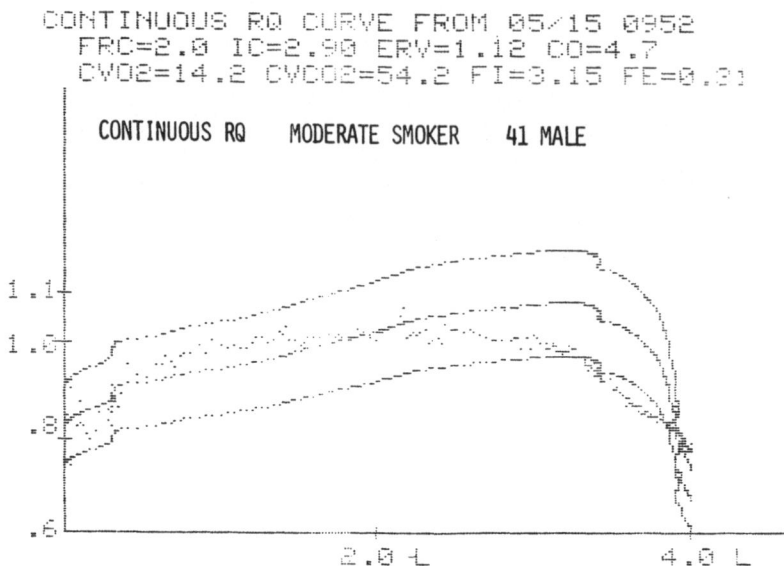

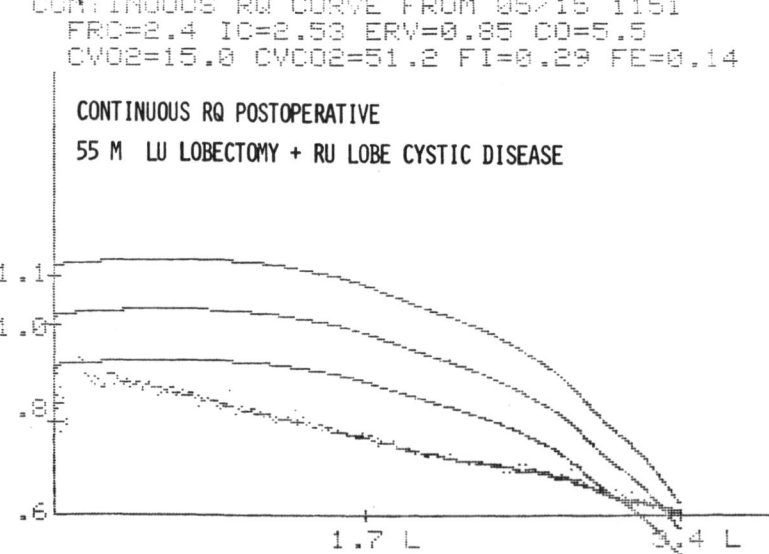

Fig. 1A. shows normal and 1B post-pulmonary resection patient who
had chronic obstructive lung disease. The model has been
found to be adequate and small variations in FRC do not
seriously distort the continuous RQ traced by the model.
With the definition of the method, we are carrying out
studies in various subsets of patients to test the continuous RQ as a measure particularly of change in shunt
fraction.

and to cardiac output (CO) and functional residual capacity (FRC). To interpret a given curve, one must use a predictive model of uniform ventilation-perfusion for the given values listed to compare with the experimental curve.

In Wagner's description of the model, he assumed that inspiration to vital capacity (VC) was instantaneous and that expiratory flow was constant and slow. We found these criteria impossible for untrained or ill subjects.

We have changed the model and checked the effects of altering its various determinants on the predicted form of the model. Our studies of the relationship between the data and the actual model, and the effects of changes in the most uncontrolled factors, FRC and CO, are presented. In order to have a test which untrained subjects and ill patients can perform, it was necessary to remove the need for an instantaneous inspiration. This requires a model which can account for varying inspiratory as well as expiratory rates and volumes. We changed Wagner's model to one which calculates the time course of dilution of FRC by inspired volume, the CO_2 added to and O_2 removed during inspiration and expiration. The model requires knowledge of FRC as a measure of initial volume, oxygen consumption ($\dot{V}O_2$), carbon dioxide output ($\dot{V}CO_2$), and CO to calculate venous oxygen ($C\dot{V}O_2$) and carbon dioxide ($C\dot{V}CO_2$) contents. We measure $\dot{V}O_2$, $\dot{V}CO_2$ and FRC, inspiratory capacity (IC), and expiratory reserve volume (ERV) before computing continuous RQ. It is difficult to be certain that FRC has been measured precisely. We use the comparison of IC and ERV during the continuous RQ manoeuvres as indicators of unchanged FRC. We assume cardiac index (CI) of 2.5 L/min unless it has been measured.

Our studies to date have examined the effects of change in inspiratory and expiratory flow rates, FRC and CO on the continuous RQ, compared to the modified Wagner model. The subjects attempted to inhale from normal FRC either at a slow constant rate as near 0.5 L/sec as possible, or as quickly as possible to VC, and then, without an inspiratory pause, to exhale at a slow constant rate as near 0.5 L/sec as possible. Following a fast inhalation, it was very difficult to prevent a short period of fast expiratory flow unless there was an inspiratory pause.

We then displayed the flow volume loop of the continuous RQ breath to characterise the flow pattern and the calculated predicted RQ curve using measured $\dot{V}O_2$, $\dot{V}CO_2$, FRC and assumed CI of 2.5 L/min to calculate CO. CO and FRC were then varied signally in interactive fashion to test the effect of errors in these values on the model. The model has been found adequate and small variations in FRC do not seriously distort the continuous RQ traced by the model. With the definition of the method, we are carrying out studies in various subsets of patients to test the continuous RQ as a measure of change in shunt fraction.

REFERENCE

1. H. J. Guy, R. A. Gaines, P. M. Hill, P. D. Wagner and J. B. West,
    Computerized non-invasive tests of lung function, A flexible
    approach using mass spectrometry, Am. Rev. Resp. Dis. 113:
    737-744 (1976).

POST-ANAESTHETIC GAS EXCHANGE: ON LINE COMPUTER METHODS

S. F. Sullivan, G. Y. Shigezawa, D. H. Huang,
R. T. Smith and S. M. Ricker

Department of Anaesthesiology, School of Medicine
University of California
Los Angeles, California 90024. USA

During emergence from inhalation anaesthesia, anaesthetic gases are eliminated from tissue stores throughout the body and body metabolism changes rapidly. In has been difficult to assess the changes in oxygen consumption ($\dot{V}O_2$), CO_2 elimination ($\dot{V}CO_2$), and respiratory change ratio (R) because of the unsteady gas state. Our approach has been to automate the standard open-circuit approach using a portable system consisting of an electronic volume displacement spirometer, minicomputer and mass spectrometer.

For example, following general inhalational anaesthesia with halothane (HAL), nitrous oxide (N_2O) and O_2, there are 6 gases in an unsteady state, three undergoing uptake (O_2, N_2, Argon) and three undergoing elimination (CO_2, HAL, N_2O). Measurement of $\dot{V}O_2$ requires that net inert gas exchange (N_2/Argon) be zero (2).

$$\dot{V}_{O_2} = V_E \left[\left(F_{I_{O_2}} \cdot \frac{F_{E_{N_2}}}{F_{I_{N_2}}} \right) - F_{E_{O_2}} \right]$$

Measurement of gas (X) elimination ($\dot{V}CO_2$, $\dot{V}HAL$, $\dot{V}N_2O$), when $F_{I_X} = 0$, is

$$\dot{V}X = \dot{V}E \cdot F_{E_X}$$

When anaesthetics are discontinued and the subject breathes air or other O_2-N_2 mixtures, measurement of $\dot{V}CO_2$, $\dot{V}HAL$, and $\dot{V}N_2O$ present no particular problem other than that of simultaneous gas analyses,

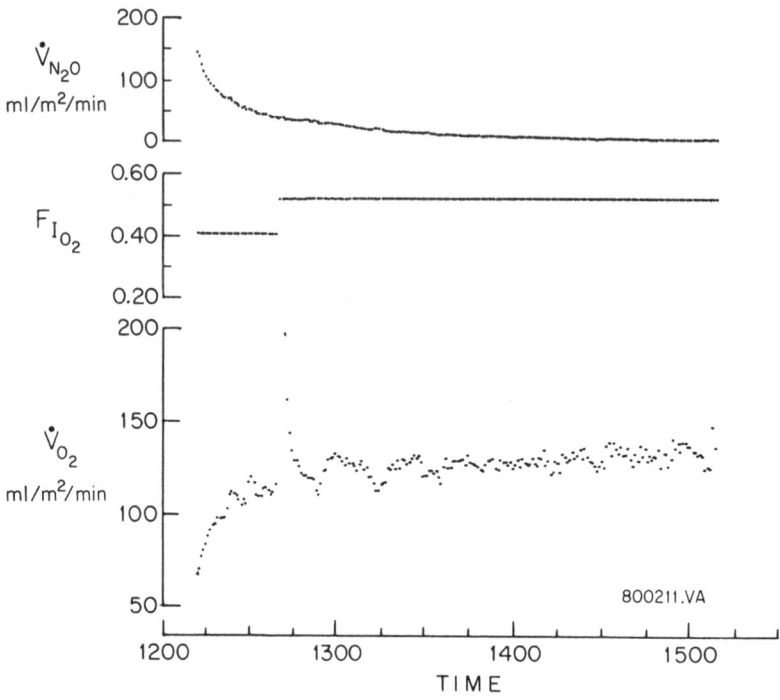

Fig. 1.

solved by use of the gas analysing mass spectrometer. Measurement
of $\dot{V}O_2$ is not immediately possible because the ratio FEN2/FIN2 is
changing as N_2 storage is proceeding. Under normal conditions the
error introduced in $\dot{V}O_2$ computation by N_2 storage is 1% at 2 minutes.
The error in $\dot{V}O_2$ measurement during N_2 wash-in or wash-out is illustrated in the figure. Fig. 1. shows results from an anaesthetised
adult whose lungs were artificially ventilated, via an endotracheal
tube, with 50% N_2O in O_2 until 1208 when the post-operative inspired
mixture becomes 41% in N_2. At approximately 1240 hours F_{IO_2} was
altered from 0.41 to 0.52. Obviously there is no effect on the N_2O
wash-out while there is a significant error in computed $\dot{V}O_2$ for
several minutes during the N_2 adjustment. This approach is ideal for
physiological study of patients in operating rooms, recovery rooms
and intensive care units. Although more difficult in the patient
without a tracheal tube, reliable gas collection is possible with a
securely fitting face mask and non-rebreathing valve. Automation of
standard open-circuit methods for measuring pulmonary gas exchange
now makes it possible to measure $\dot{V}O_2$, $\dot{V}CO_2$ and R in parallel with
quantitative anaesthetic gas elimination.

REFERENCES

1. S. F. Sullivan, R. W. Patterson, R. T. Smith and S. M. Ricker,
    Quantitative respiratory gas exchange during Halothane
    induction, in: Proc. Low Flow and Closed System Anaesthesia
    Sympos.," pp. 85-97, J. A. Aldrete, ed., Grune and Stratton
    (1977).
2. L. E. Fahri and H. Rahn, Gas stores of the body and the unsteady
    state, J. Appl. Physiol. 7:472 (1955).
3. V. A. Nowakowski, S. F. Sullivan and E. C. Deland, Oxygen uptake
    measurement error during the unsteady state, Fed. Proc. 37:866
    (1978).

CLINICAL APPLICATION OF A NEW METHOD FOR METABOLIC MONITORING IN PATIENTS AFTER CARDIAC SURGERY

E. Turner, U. Braun, K. Freiboth

Department of Anaesthesia
University of Göttingen
Germany GFR

Measuring oxygen uptake seems a logical approach to monitoring critically ill patients.[1]

A device for measuring oxygen uptake and carbon dioxide production has been used in exercise physiology for some years.[2] It is called a Metabolic Measurement Cart (MMC) - Beckman Instruments, Advanced Technology Operations, Anaheim, California - and is now applied in intensive care for the artificially ventilated patient. The aim of our investigations was to evaluate measurements of oxygen uptake during and after coronary-bypass operation with a conventional method and to compare it with the MMC measurements in the postoperative period.

METHODS

We studied a group of ten patients who underwent aorto-coronary-bypass operation. One to three bypasses were performed. Induction of anaesthesia was a standardised procedure with usual dosages of Fentanyl[R], Pancuronium[R], Etomide[R], Succinylcholine[R] for intubation and Fentanyl[R] infusion. The lungs were ventilated with oxygen in nitrous oxide to keep PCO_2 normal. Arterial, pulmonary artery (Swan-Ganz) and central venous pressure were monitored and recorded. Cardiac output (CO, Thermodilution), arterial and mixed venous blood gases were measured one hour after the induction of anaesthesia, after thoracotomy, every 15 minutes during extra-corporeal circulation, 15 minutes after extracorporeal circulation at the end of surgery in the operating room and during the post-operative period at least every hour up to five hours until the patients had reached normothermia. Oxygen uptake was calculated from CO and arterio-mixed venous oxygen content differences ($avDO_2$).

After arrival in the ICU the MMC was attached to the endotracheal tube. A more detailed description of the MMC is found in
the paper by Braun et al. The results of oxygen uptake calculated
from (A-V) PO2 difference and CO were correlated with the results
obtained from the MMC measurements.

PRINCIPAL FINDINGS AND DISCUSSION

There was a sufficient correlation between the two methods
($r = 0.837$, $n = 33$, $S_{yx} = \pm 46.8$ ml). An error of 10-20% occurs in
calculating oxygen consumption from CO and $avDO_2$. In the MMC the
error depends upon the inspired oxygen concentration, the inspiratory pressure and the moisture in the breath-by-breath line. The
alterations of cardiac index (CO/m^2BS), rectal temperature and
oxygen consumption are shown in Fig. 1.

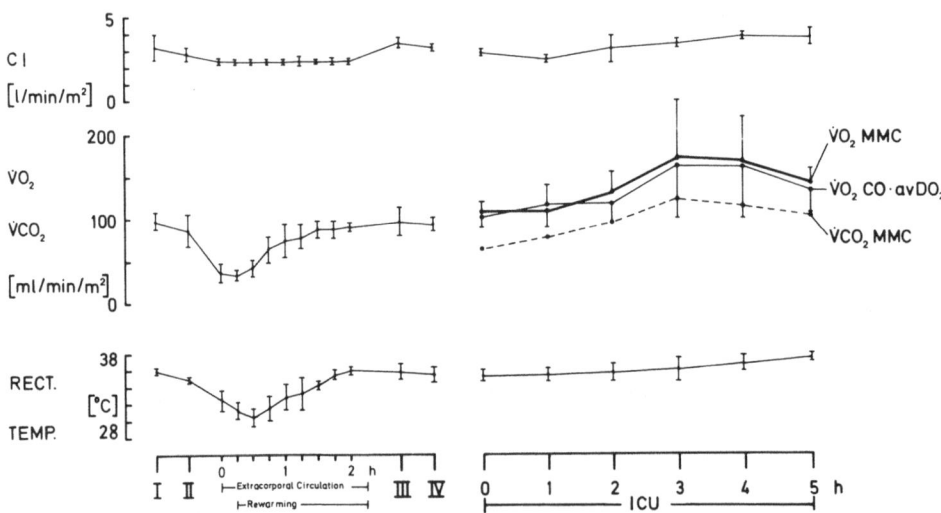

Fig. 1. Cardiac index, O_2 uptake, CO_2 production and rectal tempera-
         ture during coronary bypass operation and in the post-
         operative period.

As shown in the figure there is a sharp drop in oxygen uptake at the beginning of extra-corporeal circulation, which may be interupted as a limited shock state.[3]

As the MMC measurements are continuous more information about sudden changes in the patient's condition (e.g. shivering, hypovolaemia, cardiogenic shock) are obtained, than with conventional methods. We found a 200% increase in oxygen uptake when the patients were shivering and a marked decrease in one case of hypovolaemia. No cardiogenic shock occurred.

The highest values for oxygen consumption were found three hours after arrival in the ICU, when the patient's temperature was between 36.0 and 38.0°C. During this time shivering often occurs which explains the increasing standard deviation. MMC, CO and (A-V) PO2 calculations show the same course. Although the rectal temperature is still rising, there is a decrease in oxygen consumption after five hours, which may be explained as a stabilisation.

Future studies should include shocked patients and patients who are monitored over longer periods.

REFERENCES

1. H. Neuhof, D. Hey, E. Claser, H. Wolf, H. G. Lasch, Shocküberwachung durch kontinuierliche Registrierung der Sauerstoffaufnahme und andere Parameter, Dtsch. med. Wschr. 98:1227 (1973).
2. D. J. Wilmore, I. A. Davis, A. C. Norton, An automated system for assessing metabolic and respiratory function during exercise, Journ. Appl. Physiol. 40:619 (1976).
3. R. C. Lillehei, W. G. Manax, I. H. Block, The nature of irreversible shock, Am. Surg. 4:682 (1964).

MONITORING OF AIRWAY PRESSURE AND COMPLIANCE BEFORE AND AFTER CARDIOPULMONARY BYPASS, WITH TOTAL HEMODILUTION

O. Prakash, S. Meij and B. v.d. Borden

Thoraxcentrum, Erasmus University
Rotterdam
The Netherlands

Routine on-line measurements of airway pressure and compliance are usual parameters to monitor in the operating room during open-heart surgery. The risk of inducing pulmonary oedema has been greatly increased by the application of the total hemodilution technique. In a series of 32 consecutive patients undergoing coronary artery bypass surgery, thorough measurements were made according to the Methodological Platform described elsewhere.[1] These include hemodynamic, respiratory, and gas exchange variables. Measurements were made continuously under fentanylnitrous oxide-oxygen anaesthesia, but measurements taken before the start of surgery and following chest closure were considered as being reference points for comparison of lung mechanics, gas exchange, and hemodynamic data. From these data, which are shown in Table 1 below, it is clear that almost all patients had increased airway pressure, decreased compliance, and increases in oxygen uptake and carbon dioxide elimination. One of the patients in the series had a 70% fall in lung compliance value which was due to frank pulmonary oedema that did not respond to the administration of Lasix intravenously 1 mg/kg and Positive End Expiratory Pressure 15 cm H_2O. The other 31 patients, with increased airway pressure and decreased compliance, responded very well to this treatment within hours of closure of the chest, and were successfully extubated.

We conclude that the measurement of compliance and airway pressure are useful parameters to monitor in mechanically ventilated patients, and has proved to be a sensitive early indicator of incipient pulmonary oedema.

Table 1.

	pre-perfusion		post-perfusion		p-values	
Peak Pressure (cmH$_2$O)	17	$\pm$ 1	23	$\pm$ 1	p	0.001
Inspiratory Resistance (cmH$_2$O/l/sec)	18	$\pm$ 1	20	$\pm$ 1	p	0.001
Compliance (ml/cmH$_2$O)	57	$\pm$ 3	44	$\pm$ 2	p	0.001
VO$_2$ (ml/min/min)	110	$\pm$ 5	123	$\pm$ 5	p	0.05
CI (l/min/m)	3.6	$\pm$ 0.2	3.3	$\pm$ 0.2	n.s.	
S$\bar{v}$O$_2$ (%)	83	$\pm$ 1	74	$\pm$ 1	p	0.001
$\overline{Art}$ (mmHg)	103	$\pm$ 4	110	$\pm$ 4	n.s.	
PA (mmHg)	20	$\pm$ 1	23	$\pm$ 1	n.s.	

Values are mean $\pm$ SEM

REFERENCE

1. O. Prakash, B. Johnson, S. Meij, S. G. van der Borden, P. G. Hugenboltz, Computerized Monitoring of lung and heart function in the operating room, Computers in Critical Care and Pulmonary Medicine, 1980, Plenum Press, New York; New York 129-148